Hermann von Schrenk

A Disease of Taxodium Known as Peckiness

Also a Similar Disease of Libocedrus Decurrens

Hermann von Schrenk

A Disease of Taxodium Known as Peckiness
Also a Similar Disease of Libocedrus Decurrens

ISBN/EAN: 9783744716536

Printed in Europe, USA, Canada, Australia, Japan

Cover: Foto ©berggeist007 / pixelio.de

More available books at **www.hansebooks.com**

ASE OF TAXODIU ... NOWN AS PECKI-
ALSO A SIMILA ... BOCEDRUS DECUR-
KNOWN AS PIN-

BY

diseases o ... ttracted attention for
years befor ... made to arrive at an
anding of th ... With the investigations
Hartig, Sch ... a beginning was
but not ... searches of R.
l were ... relations between
... made clear. In
... attention to the
... w have given this

... e natural decadence
... t may be influenced
... or its functions may
... disturbing organism.

... hington University for the

... rwandlung der polycotyle-
... ngebilde, u. die daraus

... en über Bau u. Leben der

... inde des Waldes. Dresden.

... aldbäume. Berlin. 1874. —
... tc. Berlin. 1878.

1

Examples of the first class of disease need not be given; the root rot of pines * is an example of the second class, and the numerous fungi (and insects) attacking different parts of a tree come under the third class. It is particularly the latter class which Hartig has studied and with some of which the present paper is to deal.

The fungi attacking trees may be divided for convenience into such as are strictly parasitic, like the Peridermiums, Exoasci, Gymnosporangiums, etc., and such as are not. Among the latter class one finds various grades, going from the strictly parasitic to the strictly saprophytic forms, including the facultative saprophytes (of De Bary) or hemiparasites, the true saprophytes, and the facultative parasites (De Bary), or hemisaprophytes. Of the fungi which attack the trunks of trees, i. e., the wood already formed, few are strictly parasitic or hemiparasitic; the majority are hemisaprophytes, for although normally growing on dead matter they may occasionally become truly parasitic. Tubeuf † mentions a number of such fungi, among them several common in this country, such as *Trametes Pini*, *Polyporus fomentarius*, *Polyporus sulphureus*, and others. *Trametes Pini* may serve as a good example. When growing on species of pine, such as *Pinus palustris*, *Pinus Strobus*, *Pinus echinata*, it flourishes in the heartwood of these trees as a strict saprophyte, i. e., on the dead wood. The resinous contents of the living wood prevent its becoming parasitic. On the other hand, when growing on *Abies balsamea* or *Picea nigra* it becomes a parasite, growing likewise in the living wood and ultimately killing the tree. The diseases to be discussed in the following are due to hemisaprophytes, as they affect the heartwood of the trees and never enter the living parts of the trunks.

* Hartig, R. Zersetzungserscheinungen, etc. 75.
† Tubeuf, C. Freiherr von. Diseases of plants 5. (English edit. by W. G. Smith. 1897.)

Some years ago, while collecting in the cypress swamps of Arkansas, a peculiar defect of the bald cypress, *Taxodium distichum*, was noticed, popularly known as "pecky" or "peggy" cypress. Further investigation showed that the defect was prevalent wherever the cypress grew in abundance, and that fungus threads were constantly associated with the pecky wood. This led to the investigations here recorded.

In pursuing these investigations little could have been done without the generous assistance of numerous lumber companies. Among those to whom thanks are due are the Lutcher & Moore Cypress Lumber Co. of Lutcher, La.; Mr. M. E. Leming of Cape Girardeau, Mo.; Towle Bros. of Towle, Placer Co., Cal.; the Stimson Mill Co. of Ballard, Wash.; Mr. A. J. Johnson of Astoria, Oregon, and Birce & Smart of Emigrant Gap, Cal.

To Dr. W. G. Farlow and Dr. H. W. Harkness I am indebted for many suggestions; to Dr. J. J. Früh of Zürich, for his courtesy in answering some questions; to Maj. B. M. Harrod of New Orleans for assistance in obtaining buried cypress logs, and to Prof. C. R. Sanger and Dr. G. Alleman for suggestions on chemical questions. I also take pleasure in expressing thanks to Dr. Wm. Trelease for much encouragement and generous assistance.

THE DISEASE OF TAXODIUM KNOWN AS "PECKY" CYPRESS.

HISTORICAL.

The first mention of the disease of cypress known as "pecky," or "peggy" cypress, is made by Dickeson & Brown.* They say of it: "That species of decay to which it [the cypress] is most liable, shows itself in partial or detached spots at greater or less distance, but often in very close proximity to each other. It is a decomposition of the

---

* Dickeson, Montroville W., & And. Brown. On the cypress timber of Mississippi and Louisiana. (Am. Journ. of Science. ii. **5**:15. 1848.)

wood fiber to which the tops and central parts are the most
exposed, and which, when affected, appear as if operated
upon by worms. . . . Timber affected in this way is
denominated by raftsmen, '*pecky*.'"

Sargent * says of the cypress: "It is often injured,
especially west of the Mississippi river, by a species of
Daedalea not yet determined, rendering it unfit for lumber."
Farlow,† writing in Sargent's Silva, notes that "a species
of dry rot in living timber often diminishes its value, and
in Louisiana and Mississippi is said to affect at least one-
third of all the trees." Rather recently Roth ‡ mentions
its occurrence in the South, and briefly describes its appearance. Beyond these few notes, nothing appears to have
been said of the disease.

<h3 style="text-align:center">OCCURRENCE.</h3>

Taxodium distichum is now found from South Carolina
to Florida (some trees occur as far north as New Jersey §)
thence to Louisiana and northward as far as southern Indiana ¶
and southeast Missouri. Wherever the cypress grows to
any size, it shows the "pecky" disease, the prevalence of
which appears to be very variable. The exact percentage is
difficult to ascertain as it varies materially with the locality.
Roth (l. c.) says that 30% of the entire cypress supply is
damaged by this disease. As a rule one may say that
wherever the cypress grows, one will find it "pecky," and
that there are no regions where all trees are sound. As
for particular localities, Roth mentions a tract of land in
Florida, which had to be abandoned entirely on account of
"pegginess." In the Mississippi Valley by actual count it

* Sargent, C. S. Forest trees of North America. 10th Census
9 : 184. 1883.

† Sargent, C. S. Silva of North America 10 : 150. 1896.

‡ Roth, Filibert. Progress in timber physics — "Bald cypress."
(U. S. Dept. of Agr., Div. of Forestry, Circular No. 19 : 3. 1898.)

§ Hollick, A. (Cypress in N. J., read before Bot. Soc. Am. 1898.)

¶ Wright, John S. — Notes on cypress swamps in Knox Co., Indiana.
(Proc. Ind. Acad. Sci. 1897 : 172).

has been found that the trees near their northern limit are less frequently diseased than the more southern ones. As it is almost impossible to tell whether a tree be pecky or not before it is cut down, all actual counts had to be made where lumber mills were cutting the trees, and as they usually cut all trees, even those liable to be diseased, a fair estimate for that particular locality could be made. In St. James Parish, Louisiana, 397 trees out of 400 were found diseased to a greater or less extent. From circulars sent to various lumber concerns the following estimates are made, which may be considered as much under- rather than overdrawn. Apalachicola, Fla., 10–15%. New Orleans Cypress Lumber Co., 99%. Ramos, La., 15%. Georgetown, S.C., "considerable." These figures refer to "peckiness" in logs used for lumber, and do not have any reference to the tops of trees, which are the first parts to be "pecky." The character of the ground seems to have little if any effect on the prevalence or extent of the disease. The cypress trees normally grow with their root system in water for at least a part of each year, and in many places, particularly along the coast, during the entire year. This rather unusual habit of growth together with the appearance of the puzzling formation of knees has led many to connect the facts of growing in water, development of knees, and "peckiness." So far no evidence is forthcoming to show any connection between these factors.

NAME.

When a diseased cypress tree is cut down, the heart wood appears as if a large number of holes had been bored with a $\frac{1}{4}$ inch bit which had been withdrawn, leaving the shavings, finely divided, within the hole. It is this peculiar appearance which has given rise to the different popular terms applied to the disease. Dickeson & Brown (l. c.) refer to it as "pecky." In the Mississippi Valley and throughout Louisiana I have found the diseased wood called "pecky" cypress, and the disease itself called the " peck." In North Carolina the term " botty " (see Roth l. c.) is more or less common because of the supposed action of a larva, the " bot." "Peggy " is frequently used in Georgia and Florida, where correspondents also give the term " puck." Near pinelands " punk " is used by pine lumbermen, accustomed to the decay caused by *Trametes Pini*. It is almost useless to speculate as to the origin of the various terms, and a choice between them is difficult. Having found the term " pecky " most widely known as well as the one which was first used, I shall call the disease by that name throughout this paper.

APPEARANCE OF WOOD.

The diseased wood appears full of holes (Pl. 1, fig. 2), varying in width from $\frac{1}{4}$–$\frac{3}{4}$ inches. These holes are found in the heartwood only, and in trees after they have reached the age of 125 years or thereabouts. Young trees of *Taxodium* are comparatively rare, but such as were noted, varying in age from 50 to 125 years, were always free from any defect. The holes in the wood extend longitudinally up and down in the trunk, parallel to the wood fibers. The holes never extend transversely. They are separated from one another by layers of wood apparently perfectly sound. They vary in length from $\frac{1}{4}$ inch to 6 inches, or longer in some cases; most frequently they are 4–5 inches long. They end bluntly at both ends, and as a rule do not communicate. Frequently trees are found in which some holes do open into one another, but these are rather exceptional. The holes are filled with a yellow brown powder which readily crumbles into the finest dust between the fingers. The powdery mass does not completely fill the space, showing that much material has been destroyed. Occasionally the mass is not entirely composed of the powdery substance; stringy fibers, composed of wood cells not yet disintegrated, fill the cavity, together with much finely divided matter. This indicates that the disintegrat-

6

ing factor has not acted uniformly but has caused certain parts to decay, while others are spared. A fluffy white fungus mycelium, covered with drops of liquid as with dew, is oftentimes present, but more often none is to be seen amid the dry contents of the holes. Several trees

APPEARANCE OF WOOD.

were found near Mobile, Ala., in which the holes were partially filled with a peculiar reddish-brown, soft substance having a bright shining fracture. This substance— adhered firmly to the walls, as if forming a part of the wood fibers. It will be described more in detail. Associated with this substance peculiar white needle-shaped

7

crystals were found whose identity has not yet been determined, as the amount found was too small to admit of analysis. This substance had the following properties: very light, like fine cotton wool, or cocoon silk, apparently very pure; volatilizes at once on platinum (heated) without burning; insoluble in water, soluble in hot alcohol, from which it crystallizes in shapes looking like sea moss; very soluble in petroleum ether, and extremely so in chloroform; residue colloid, resinous; melting point 174° C., pretty sharp without decomposition; chloroform solution does not absorb bromine; sublimes very readily, forming beautiful hairlike crystals.

A number of trees were found in which the holes, instead of being filled as stated above, were nearly empty. They had a shining white lining (Pl. 4, fig. 3) from which isolated white fibers projected into the cavity. The white fibers were found to be pure cellulose.

When the brown contents are brushed out of the holes a perfectly even and smooth surface is left on all sides, indicating a very sharp dividing line between the decayed elements and those apparently sound. A board from which the powder has been taken looks as if a number of grooves had been cut with a gouge chisel (Pl. 6).

In a tree the peckiness starts in the upper part, i. e., the majority of the trees are perfectly sound at the base, and very much diseased in the upper portion of the trunk and the larger branches. The decay may extend but a few inches up and down, or for several feet, or through the entire length of the tree. The youngest branches in which any peckiness was found were 60 years old. Radially it may appear over the entire cross-section or on but one side. It is by no means the rule that the innermost rings are the first ones to decay (Pl. 1) as might be supposed from analogy with other timber diseases. A large tree at Arbor, Mo., approximately 300 years old, was pecky to within 25 ft. of the base, another to within 35 ft. The

8

peckiness in the latter extended upward into two main branches for 20 ft. At the points where all recognizable traces of the disease ceased, the branches had about 150 annual rings. A third tree was sound for 60 ft. from the base, then became very pecky, the peckiness passing up into two or three main branches, and still another was pecky 3 ft. from the base and upward. The extent of peckiness varies, i. e., in one trunk the holes may be several inches apart, or scattered all over the cross-section, in another they may be confined to the first 150 rings. Nowhere was a single tree seen hollow, at least none which was hollow because of an advanced stage of peckiness. In this respect this disease affects the cypress just as *Trametes Pini* does the pines. One never finds a pine hollow because of disintegration caused by *Trametes Pini*. This is especially to be noted, as it will be referred to again. It is a noticeable fact that in traveling through cypress forests one rarely sees many fallen trees, and where violent windstorms have overthrown any number of trees, these are just as often trees sound at the base as those which are diseased. In the fall the whole tree falls, i. e., the trunk does not break, as do many pines and deciduous trees, in which the entire heartwood may have been destroyed, by such a parasite as *Polyporus sulphureus*. The oldest trees e. g., in which about 1800 rings were counted (southern Louisiana), have the same appearance as those but 200–300 years in age.

STRUCTURE OF DISEASED WOOD.

The wood of *Taxodium* is composed of tracheids with one or two rows of pits. The growth rings are rather broad, the summer wood about one-third the width of the spring wood. Resin passages are wholly wanting, and in their stead there are numerous resin cells either scattering or in tangential bands.* The amount of resin in the wood is

---

* Penhallow, D. P. The generic characters of the N. A. Taxaceae and Coniferae. (Trans. Roy. Soc. Canada ii. 2 : 51. 1896.)

comparatively small. The diseased wood is darker in color than the normal wood, has no tenacity, and when crushed, turns into fine powder. These properties lead one to suspect profound morphological as well as chemical changes. A radial section through a 'rotted hole and the adjacent sound wood is represented on Pl. 5, fig. 10. The normal wood cells (a) show the constituent lamellae of the cell-wall plainly. The first noticeable change is in the bordered pits, which look as if they were corroded, like starch-grains in process of solution. This appearance is due to drops of resinous oil, which arrange themselves in this peculiar manner (Pl. 5, fig. 3) on the walls and within the cavity of the pit. When treated with turpentine the resin is dissolved and the pits then look smooth. The walls at the same time are perforated in many places by colorless hyphae; no preference is shown for the pits. As one proceeds toward the decayed spot the pits look fragmented; finally peculiar spiral breaks appear extending from the pits and passing upward from left to right. The breaks of two adjacent walls cross one another at the pits (e), the lower one apparently extending from right to left. The whole cell-wall becomes striated, the striae all extending in a spiral line from left to right around the wall. Before long the circular disc of the pit drops out, leaving a hole with jagged outline, which gradually increases in size. The number of breaks in the walls has increased, and finally the whole wall breaks up into innumerable pieces. The wood, when once the disintegration sets in, becomes so very brittle that it is very difficult to get good sections. Imbedding the same in soft paraffine was found very useful. The longitudinal walls grow thinner because of the shrinkage of the middle lamella. This gradually disappears, and gives rise to the breaks in the wall already spoken of. The shrinkage and solution finally has gone so far that only the primary lamella is left, which breaks into many pieces. The dropping out of the more resistant walls lining the pits is char-

10

acteristic, and in preparations of much-decayed wood large numbers of the circular discs can be seen floating about.

The micro-chemical reactions are marked. Any investigation into the chemical nature of wood substance is apt to be rather unsatisfactory. It is possible to record certain well-marked reactions, but often their true significance will not be apparent, because our knowledge of the complex constituents of wood, and particularly of its decomposition products, is still so very meager.

The most characteristic reaction is the one with phloroglucin and HCl. If a section, preferably a transection of wood cut so as to include the outer portion of the decayed area, and some of the surrounding wood, be treated with this reagent, an appearance such as is represented on Pl. 3, fig. 1, is obtained. The cells of the sound wood i. e., wood in which no recognizable morphological change has taken place, stain dark red purple. The primary lamella stains much deeper (p). Passing to cells further inward (towards the diseased spot) the tertiary lamellae of some cells no longer stain red purple but yellow (d). This yellow coloration increases as one passes on, the red decreasing correspondingly, until at a certain stage only the primary lamella is stained red. The pits are the first areas to show the yellow color. On a radial longitudinal section the contrast between the surrounding wall and the pit is very marked, the latter looking like a hole in a red field. In the final stage the remaining parts are entirely yellow, no red being visible. The yellow coloration appears first along the medullary rays, and is always in advance of the same reaction in the intervening wood cells. Hand in hand with the disappearance of the red color goes the shrinkage of the secondary lamella, as described. This reaction of the cell-wall is due to the gradual extraction of the coniferin elements of the walls. They are at first extracted from the innermost lamella, then from the secondary lamella, and last of all from the primary

11

lamella and the intercellular substance at the angles
of the cells. If a similar section is treated with chlor-
iodide of zinc the walls of sound wood cells are colored yellow-brown. The cells from which the lignin elements have been removed stain brown likewise. This
indicates that they are not cellulose. In this respect the
disintegration of the cypress wood differs from that caused
in wood of the yellow pine by *Trametes Pini*. In the latter
there is one form of disintegration in which the lignin elements are gradually removed from the cell-wall, beginning
with the secondary lamella, closely followed by the tertiary
lamella. After this extraction a much thinner wall of pure
cellulose remains. Some cases were found in wood of
Pinus echinata, however, which could not be distinguished
from pecky cypress, i. e., after the extraction of the lignin
elements, as indicated by phloroglucin and hydrochloric
acid, a membrane remains which is not cellulose. Hartig*
describes a reaction much like the foregoing one in pine
wood attacked by *Merulius lachrymans*, of which he says:
" It appears as if there were a certain relation between the
coniferin content of the cell-wall and the ease with which
the wood is destroyed." This test is so delicate that it
shows the presence of a disturbing cause in the wood
long before any evidence can be detected by the micro-
scope. Other lignin reagents give similar results, although
not so striking. Aniline sulfate turns sound wood brilliant
yellow, while it leaves the affected lamellae almost colorless. Thallin and phenol give similar reactions. If the
sections are treated with dilute KOH the normal wood
cells are not affected beyond very slight swelling. The
diseased cells swell more or less, particularly those parts
which stained yellow with phloroglucin. After prolonged
action of KOH, the delignified parts stain blue with chlor-
iodide of zinc. This indicates that the first change in the

* Hartig, R. Der ächte Hausschwamm 53. Berlin. 1885.

cell-wall is the removal of some of the incrusting substances, probably coniferin and vanillin. The KOH re-
moves the remaining substances and leaves the cellulose
membrane free to react with chloriodide of zinc. Very
much rotted wood stains intensely blue after treatment with
KOH. The blue color appears first about the pits, and
diffuses towards all sides, looking much like an inkspot on
which water has been dropped, causing it to diffuse irregularly over the surrounding area. Sound wood stains yellow-
brown with chloriodide of zinc, even after treatment with
KOH. With iodine and sulphuric acid, rotted wood stains
brown.

A very different form of disintegration now and then
occurs (Pl. 4, fig. 3). What the reason is why one form
occurs at one time and a second at some other time, I cannot explain. Large holes appear in the sound wood, filled
with a spongy mass of white fibers. The holes have a
white or tawny lining of fibers, which can be pulled off in
groups. These holes are as large as the ones filled with
brown powder. The change in the wood cells is almost
exactly like that found in pine wood attacked by *Trametes
Pini*,* above described. The secondary lamella is gradually
changed so that it stains purple with chloriodide of
zinc, that is, the lignin substances have been entirely
removed. Very soon after the first signs of delignification
become evident, the primary lamella separates into two
lamellae, which are then dissolved. This causes the indi-
vidual wood cells, or rather the cellulose fibers, to fall
apart. No intermediate steps between the lignified cell-
wall and the cellulose wall are to be detected, which gives
the impression that the extraction of the lignin elements
must take place all at once. The amount of pure cellulose
fiber thus found in one hole is surprising. From a hole
3–4 inches long several grams were obtained of many white

---

* Hartig, R. Zersetzungserscheinungen, etc. 35.

fibers without any impurities whatever. The quantity of wood transformed into cellulose is exceeded only, as far as I know, by that found in diseased wood of *Juniperus*, decayed by *Trametes Pini*, to be described in another paper.

With polarized light, the prisms being crossed, the primary lamella of sound wood appears white, i. e., it is highly refractive; the secondary lamellae are darker. The rotted wood, with the exception of some very minute particles, allows no light to pass. The hypothesis of a crystalline structure of the cell-wall, as advocated by Nägeli, is based largely on its optical properties. Nägeli held that the double refraction indicated a condition of stress in more than one direction. The absence of any refraction in the rotted wood indicates a homogeneous condition, i. e., one in which the stress is equal in all directions. The change from sound wood to the decayed form must have been a profound one to bring about this condition. It has been noted for wood destroyed by *Merulius lachrymans*,* that it separates the white polarized light into blue and yellow parts. Hartig makes no attempt to explain why this should be so. In this connection it may be said that rotted wood of *Libocedrus decurrens*, yet to be described, and wood of *Juniperus Viginiana* destroyed by *Polyporus carneus*, appear dark when viewed with crossed prisms.

HUMUS COMPOUND.

In the cells immediately surrounding the rotted areas certain parts of the walls are colored dark brown by an apparently homogeneous substance. This occurs in various forms. Most commonly it has numerous cracks and fissures breaking it into many plates, looking much like mud which has dried in the sun (Pl. 4, fig. 4), then again it appears in the form of irregular granules scattered along the walls,

---

* Hartig, R. Der ächte Hausschwamm 61. Berlin. 1885.

14

usually more numerous at the lower-end of the cells. When KOH is added the whole mass dissolves slowly, melting away like wax; the tracheids become filled with the red-brown solution. By treating finely divided wood with dilute KOH this substance can be extracted in quantity. If the potash solution is neutralized with dilute HCl, a red-brown flocculent gelatinous precipitate is formed slowly, which gradually settles to the bottom. When dried it resumes the appearance seen in the tracheids. In mass it is reddish-brown, soft, tasteless and odorless, insoluble in alcohol, ether, chloroform, acetone, turpentine, etc., but very soluble in alkalies, KOH, $NaHPO_4$, etc., and can be reprecipitated from such solutions by acids. Because of its peculiar physical and chemical properties the substance is classed among the humus compounds. This humus compound, as it will be designated, was evidently at first in a liquid condition, as it fills the cells so evenly. Furthermore, wherever mycelium occurs, this is coated with a layer of the compound, so that the walls of the hyphae look brown and show several contour-lines. Wherever there is any sign of decomposition, there this product appears immediately. It is at first seen in the medullary rays, filling the cells and obscuring their contents so that nothing can be distinguished in the cells. The brown contents of the rays extend out through many annual rings from the initial point of decay. On several occasions trees were found with exceptionally large quantities of this material. In these trees the cavities or holes had a brown powdery mass lying loosely within, but the bounding walls had a thick coating of the brown substance. It was quite soft, broke readily with a shining fracture, was non-elastic and dissolved readily in alkalies, in fact agreed so closely with substance already found in the tracheids as to leave no doubt as to their being one and the same substance. The finding was of value as it was possible to trace the origin of this compound directly, which was not

15

possible in the majority of cases. Our present knowledge of the humus compounds is at best a meager one. They are generally described * as black bodies, which form in the decay of organic substances, and which occur in soil, peat, etc. They are divided † into three groups (according to their solubilities): 1. Such as are soluble in alcohol and dilute alkalies. 2. Such as are very soluble in alkalies and precipitated by acids as gelatinous bodies insoluble in alcohol. 3. Such as are very soluble in alkalies, precipitated by acids, the precipitate soluble in alcohol. The substance found in the cypress wood belongs evidently to the second class, one to which a large number of products belong, particularly those obtained from peat and decaying vegetable substances.‡

Much has been written on the humus compounds, particularly those found in peat. Mulder,§ Hoppe-Seyler,‖ Griesebach,¶ Senft,** Früh,†† have treated more or less of various compounds. Früh' gives the best general account and the following notes are taken from his paper. He says (p. 63): Ulmates and humates, ulmin and humin, ulmic and humic acids in homogeneous masses or in fine particles give a mass which when moist is slightly elastic. In drying these substances contract, become black, shining like glass, hard, and break, with conchoidal fracture. The splinters

---

* Beilstein, F. Handbuch der organischen Chemie 1 : 1107. 1893.

† Hoppe-Seyler. Hoppe-Seyler's Zeit. f. phys. Chemie 13 : 1101.

‡ Some of the humus compound was sent Dr. Früh who says of it: "It seems to agree in its properties with ulmic acid, or a calcium salt of the same." Dr. Van Bemmelen of Leiden has kindly undertaken to make a more detailed examination.

§ Mulder, Liebig's Annalen der Chemie (u. Pharmacie) 36 : 343. 1840.

‖ Hoppe-Seyler. 1. c.

¶ Griesebach. Über die Bildung des Torfes in den Emsmooren. Göttingen. 1846.

** Senft. Die Humus, Marsch und Torfbildungen. Leipzig. 1862.

†† Früh, J. J. Über Torf und Dopplerit. Zürich. 1883. (Gives long bibliography.)

are yellow-brown at the edges, transparent and soluble in 5% KOH in the form of the acid or as ulmate or humate; humin and ulmin simply swell in 5% KOH. "In the humification yellow-brown places appear on the cell membrane, which can be bleached out with KOH, in the form of ulmin or humate; the remaining cell membrane shows distinct cellulose reaction. Lignified membranes ulmify with difficulty, although wood cells can change completely into peat." Griesebach (1. c.) mentions the transformation of wood of *Erica Tetralix* and *Calluna vulgaris* into ulmin substances. Humic acids have been found in plants, and Früh mentions a number of cases. Thus according to Lucas, Einhof extracted the same from spores of *Agaricus atramentarius*, while he himself obtained one from *Uredo segetum*. Früh isolated humic acid from spores of *Elaphomyces granulatus*.

In the wood of *Taxodium* in which the large masses of humus compound were found the transition from lignin to the humus bodies was very evident. Pl. 3, fig. 2, represents a section made through the border of a hole, after staining with phloroglucin and HCl. At "g" the primary lamella is seen, dark red, indicating the presence of coniferin, etc. In the next row of cells the interior is coated with yellow-brown masses (h) which in the unstained wood contrast beautifully with the almost white cellwall. These masses are found to be humus substance, readily dissolved by dilute KOH. The phloroglucin stains the secondary lamella. Between this normally lignified portion and the inner humus layer is a layer staining yellow. This is evidently similar to the membrane already described (Pl. 3, fig. 1 " d "), i. e. the wood substance gives neither a lignin nor a cellulose reaction. After treatment with KOH it stains deep blue. This is in part the process as described by Früh, except that here there is an intervening step between the lignin and the humus substance. Passing now from the cells just described, one finds the layer

17

of humus increasing in width (c). In drying, numerous fissures have appeared in the mass. The lignin layer becomes narrower and narrower, then disappears and at last even the primary lamella no longer gives the lignin reaction, and the whole is transformed into humus compound (e). The positions of the original cells are still very evident, and here and there a piece (u) of unchanged cell-wall remains in the homogeneous mass of matter.

The action of the rotted membranes on polarized light has already been mentioned. The primary lamella shows decided light lines in a transection of the kind shown on Pl. 3, fig. 2, but as soon as the wood no longer gives the lignin reaction it appears dark when the Nicol prisms are crossed. The same is true of the humus compound. Whatever the change is which changes a non-homogeneous body to a homogeneous one, it is one which takes place when the chemical structure of the non-homogeneous body begins to change. When a portion of the humus mass is dissolved in dilute KOH there appear in the center of this mass certain highly refractive bodies $\frac{1}{2}$–$1\frac{1}{2}$ μ in diameter, of very definite structure resembling human blood corpuscles somewhat (Pl. 5, fig. 8). They are hexagonal in shape with blunted corners and have a much depressed center, so that the edge view shows four contour lines, two parallel lines, and two of an hour-glass shape. When examined with polarized light they shine brightly when viewed from the edge, and as they have a decided Brownian motion, they alternately flash and disappear. Their very variable size, but constant form, as well as their appearance in polarized light, suggest that they are crystals of some sort. Hartig * says that the comparatively high resistance of the walls bounding the lens-shaped pits, is probably due to the large number of calcium oxalate crystals imbedded in these walls; he indicates these by fine dots (fig. 13). The

---

* Hartig, R. Der ächte Hausschwamm 57

small bodies which he finds he describes as "rounded in form." He suggests that they may be crystalline, and that the deposition of the lime in the walls is made in the form of crystals. "If this be true it may be asked whether the action of the cell-wall when viewed with polarized light may not be explained by the refraction of these bodies." Hartig, however, did not prove the crystalline nature of the bodies. The small bodies from the humus compound dissolve in HCl, which might indicate them to be crystals of calcium oxalate. They are evidently massed together in the humus compound, and become visible only when the latter is removed. It is not at all improbable that they were constituents of the cell-wall, which were not destroyed by the disintegrating factor and remained unharmed, imbedded in the liquid mass of humus compound. The present data do not warrant any definite conclusions as to their real nature and origin.

From the description just given of the formation of the humus compound, and comparing this with the normal method of disintegration of the wood, it seems that the process may be summed up as follows: For some reason, the normal lignified membrane changes, i. e., certain of its constituents, which ordinarily react with phloroglucin, are extracted. Then more profound changes take place ending in the formation of a humus compound. This ordinarily diffuses through the adjoining cells, and ultimately hardens in the tracheids surrounding the rotted area and in the medullary rays. At the same time all contents of the cells, hyphae, starch grains, etc., are covered. Numerous experiments were made to determine the approximate per cent. of matter soluble in dilute KOH, both in much rotted wood and in wood immediately surrounding the holes, apparently sound. The amounts were found to vary between wide limits. On an average about 34% was obtained from much rotted wood, the remaining 66% consisting of pieces of wood fibres not transformed. In the wood immediately surrounding

19

the holes 6–8% was obtained. The soluble matter was precipitated from the 2% KOH solution with dilute HCl and dried at 100° C.

WOOD BETWEEN THE ROTTED AREAS.

The wood between the holes is darker in color than the normal wood, but cannot be distinguished from it structurally. Numerous fungus threads pass through the walls or have punctured them in many places. Near the holes much of the humus compound occurs, and many of the pits show the peculiar arrangement of oil globules. The specific gravity of sound heart wood and that of the wood between the holes, was determined by weighing blocks and measuring them. As the plates of wood between the holes as a rule are but $\frac{1}{2}$-$\frac{3}{4}$ inch wide, and the mass of wood not occupied by holes but $\frac{1}{4}$ inch long, the pieces to be measured had to be rather small. To bring the two tests under similar conditions, the blocks from normal wood were made of similar size. The blocks were dried at 100° C. until approximately constant weight was reached. The specific gravity of sound wood was found to be .508; that of the other, .401. These figures are probably only relatively correct, but as each is the average of a number of blocks, they seem to show that even if no visible change has taken place in the wood between the holes, some change must have occurred, otherwise there would not have been so great a difference in specific gravity. Very pecky cypress planks which had been exposed in lumber yards for many years, were examined. The powder and wood fibers which had filled the holes had been washed out and had left a smooth, even surface. The wood was to all intents and purposes very sound, and no change except numerous perforations in the walls, and the presence of much humus compound, could be detected.

It is this property of the pecky cypress not to pass beyond a certain stage of decay, which has made it possible for the wood to be utilized in a variety of ways. Dickeson & Brown call attention to this fact: " There is this peculiarity of this disease, that the cutting down of the timber arrests its further progress, and timber thus affected, although not as strong, is found to last as long as that which is very sound." This is probably a unique case of specifically " rotten " wood still capable of being used for commercial purposes. The durability of cypress timber is universally admitted, and pecky cypress does not seem to be much less so. Where it is, as in this case, a question of dollars and cents, the testimony of practical lumbermen is especially valuable. Thus, whereas sound cypress lumber sells for $20–$25 per 1,000 ft. B.M., pecky cypress sells for $5–$10 per 1,000 ft. B.M.; generally from $5–$8. One firm makes two grades of pecky planks: " pecky," and " dangerously pecky; " " the latter means that the holes are so large that a mule might put his foot through "! Mr. G. M. Bowie, of Whitecastle, La., writes: " I am watching some pecky planks laid on the ground, exposed to rain and sun; they are unchanged so far in ten years." Throughout the Southern States pecky cypress boards are used for bridge planking on plantations, for siding, sidewalks, flooring, culverts, foundations under brick work, in wet places, etc. Mr. A. S. Mohr, of Apalachicola, Fla., says: " We use pecky planks 2 in. thick and upwards for making driveways, wharfs, tramways, and for such purposes it is invaluable."

In Mobile, for instance, where there are open ditches along many streets, the vertical bank, flanking the pavement, and the bottom of the ditch are lined with pecky boards, and their lasting powers seem to be fully equal to sound boards. (Pl. 6.)

For sidewalks, such lumber is used in almost every town or city within the reach of cypress swamps, and when the

21

softer powder has worn away the grooved boards have a singular appearance.

From such data it may safely be said that the disintegration never goes beyond a certain stage. When a tree is cut down the further progress of the disease is stopped. No tree, as far as is known, has been seen in which all the wood had been destroyed, and it is for this reason that a diseased tree remains standing even when much decayed.

STRENGTH OF CYPRESS WOOD.

A number of tests were made to determine the relative crushing strength of sound cypress wood and that of very pecky wood.* In making these tests blocks cut from the heartwood were used. These were dried in a kiln for three days and were tested immediately after being taken from the drying oven. The tests were made with the machinery used for the timber tests of the U. S. Division of Forestry.† A full description of the same will be found in the bulletin referred to.

CRUSHING STRENGTH (ENDWISE), OF SOUND CYPRESS. (HEARTWOOD.)

No.	Dimensions of block. Inches.	Height. Inches.	Area. Square inches.	Breaking load. Pounds.	Breaking load per square inch. Pounds.	Locality.
1	1.32×1.30	3.24	1.71	12,300	7,191	From Lutcher, La.
2	1.69×1.27	3.23	2.14	15,490	7,144	" " "
3	1.55×1.29	3.00	1.99	14,250	7,160	" " "

---

* In making these tests I am much indebted to Mr. W. H. Henby for material assistance.

† Timber physics. Pt. I. (Bull. U. S. Div. of Forestry 6: 31.) Washington. 1892.

CRUSHING STRENGTH (ENDWISE), OF PECKY CYPRESS. (HEARTWOOD.)

No.	Dimensions. Inches.	Height. Inches.	Area. Sq. inches	Breaking load. Pounds.	Breaking load per square inch. Pounds.	Character.
1	2.13×3.25	5.98	7.14	35,170	4,925	A little sapwood in one corner. Upper surface showed 4 pecky spots.
2	2.65×3.17	6.87	8.40	43,680	5,200	Near center of tree, 14 holes visible on upper surface.
3	2.92×3.16	6.86	9.28	40,130	4,350	*Very* pecky. 14–18 holes visible on upper and lower surfaces.
4	3.25×3.37	6.55	10.95	56,050	5,120	*Very* pecky. Same appearance as No. 3.
5	3.47×3.52	7.75	12.21	40,850	3,345	*Extremely* pecky.

CRUSHING STRENGTH (ENDWISE), OF PECKY CYPRESS BEFORE DRYING. (HEARTWOOD.)

No.	Dimensions. Inches.	Height. Inches.	Area. Sq. inches	Breaking load. Pounds.	Breaking load per sq. inch. Pounds.	Weight moist. Grams.	Weight dry. Grams.	% Moisture.
1	3.03×3.22	4.37	9.75	30,200	3,097	335	278	17.0
2	3.30×3.44	3.06	11.35	35,680	3,143	275	238	13.6
3	3.51×3.55	3.29	12.46	35,970	2,886	287	261	9.0

In these tables but a few tests are given: a more extensive list is in preparation. A glance at these tables will show how comparatively strong the very pecky wood is, which is a rather surprising feature. It is of course diffi-

23

cult to state exactly how pecky a certain piece is, but the samples tested were considered as fair averages of the grade generally used for sidewalks, etc. The block marked 5 was very much more pecky than the others. From the third table it appears that when wet the wood is less strong than when thoroughly dry, which is true of all woods. The breaking of the pecky blocks was almost without reference to the holes. The wood between the holes had to stand the load, and that it was capable of holding up as much as it did is another proof of its comparative soundness. The number of tests made so far is as yet too small to determine whether any relation exists between the absolute weight of wood fiber present in the pecky logs and the breaking strength.

Within the holes, and throughout the heartwood of a diseased tree, the mycelium of some fungus is constantly met with. This is present but sparingly, and rarely forms extended masses or felts. In spite of extensive and searching examinations of a very large number of cypress trees for several years, no fruiting organ has yet been met with. The only fungus ever reported was the one mentioned by Sargent * which, as far as can be determined now, had little adequate foundation. Of the other fungi hitherto reported as growing on *Taxodium* none could be brought into any causal connection with the mycelium always found in the tree. It is to be hoped that before long the fruiting form may be discovered.

The disintegration of the wood is, in many respects, like that brought about by *Trametes Pini,* but so far there is no evidence to prove that this is the fungus which causes the "peckiness."

Wherever there is any sign of decay in the cypress wood,

---

* Sargent, C. S. Forest trees of N. A. 10th Census 9 : 184.
24

a distinct mycelium is present, and the probabilities are strong that it is the one which brings about the decay of the wood. The hyphae are brownish when young, but soon become colorless. Their chief and striking characteristic is the presence of very many clamp connections (Pl. 5, fig. 1). Brefeld,* Hartig,† and others have shown that these organs are to be found among the Basidiomycetes, particularly among the *Agaricineae* and *Polyporei*. Brefeld (l. c.) describes their formation in *Coprinus stercorarius* as lateral outgrowths of one cell fusing with the cell beneath it, and then forming a separating wall. At such points numerous branches usually appeared. Hartig (l. c.) describes the clamp connections of *Merulius lachrymans*. In this fungus they bud out, and form a branch, sometimes before the separating wall in the clamp has fused with the next cell. This is a unique case among the Hymenomycetes, as the clamps are '' sterile '' in all other forms.

The clamp connections occur on all parts of the mycelium found in *Taxodium*, but in no case did any of them branch as they do in *Merulius*. The mycelium consists of large hyphae with distinct thin walls, and hyphae of smaller diameter. The larger hyphae are constricted at the points where two cells join. They branch frequently, giving rise to the hyphae of smaller diameter. These in turn branch and rebranch. At certain points a short branch is given off which divides very rapidly into the finest threads, of hairlike dimensions. These smaller hyphae penetrate the cellwalls in all directions. Connections between adjacent hyphae occur frequently, also complicated masses, where large numbers of hyphae have fused more or less. As a rule there is but very little mycelium to be found either in the much rotted wood or the intermediate parts. Numerous holes occur all through the wood, indicating where

---

* Brefeld, O. Untersuchungen über Schimmelpilze 3 : 16. 1877.
† Hartig, R. Der ächte Hausschwamm 14. *pl. 1, fig. 3.*

the hyphae had passed through the walls. The threads show no preference for the pits. The holes often have the shape of a figure 8, i. e. they enlarge within the secondary lamella, a feature which is common to wood destroyed by many wood-destroying fungi. The scarcity of mycelium is striking, resembling in this respect wood in which *Polyporus sulphureus* has been growing.* In branches where the disease is in its youngest stage, the mycelium occurs more plentifully in those areas, which correspond to the holes to be formed later on. Between these areas the hyphae pass, boring through the tangential walls.

Besides the colorless mycelium, a mycelium is often present in the wood between the decayed holes. This appears to belong to some saprophyte, which has nothing to do with the original decomposition. This mycelium is composed of brown threads which pass through the tangential walls preferably and follow the direction of the tracheids up and down. These hyphae form marked attachment organs when boring through the cell-walls. Frank† described such organs as formed by the germ tubes of *Fusicladium tremulae.* The hypha when it touches the epidermis forms a swelling with one or more pores, from which fine tubes push through the walls into the epidermal cells. He called the swollen parts "Appressorien" or attachment organs, and believed that they aided the hypha in puncturing the wall. De Bary ‡ found similar organs in germ tubes of *Peziza Sclerotiorum;* these were formed "owing to a mechanical stimulus, which the resistance of a solid body exerts on the hyphal branches." Büsgen § described the

* Hartig, R. Zersetzungserscheinungen des Holzes 110.

† Frank, B. Über einige neue u. weniger bekannte Pflanzenkrankheiten. (Ber. d. deut. bot. Ges. **1** : 30. 1883.)

‡ De Bary, A. Über einige Sclerotinien u. Sclerotienkrankheiten. (Bot. Zeit. **44** : 377. 1886.)

§ Büsgen, M. Über einige Eigenschaften der Keimlinge parasitische Pilze. (Bot. Zg. **51** : 53. 1893.)

attachment organs as smaller portions of a hypha, formed as a result of contact or pressure irritation. " These organs adhere very closely to the cell-walls and in that way probably act as braces to give the penetrating hypha an opportunity to exert the mechanical pressure necessary to penetrate the cell-wall." Miyoshi * found that in order to penetrate a wall " the fixation of the hypha was absolutely essential " and explains the formation of attachment organs as tending in that direction. Hartig † figures several cases of swollen hyphae in the mycelium of *Polyporus vaporarius*.

The penetration of the cell-wall is brought about, according to Brefeld,‡ Büsgen (l. c.), Miyoshi (l. c.), Ward, § and others by the chemical action of a ferment given off by the tip of the hypha, aided by pressure. In the diseased wood of *Taxodium* the brown hyphae pass through the cell-walls of the wood fibers of both spring and summer wood in a radial direction. The path of a hypha is made up of a succession of short curves, each within the lumen of a wood cell (Pl. 4, fig. 1). When the tip of a hypha touches the wall it is deflected considerably, as if the hypha were pressing against the wall and pushing along the same. At the same time the tip swells, and a thread of much smaller diameter pushes into the wall. Sometimes there may be two such threads (Pl. 5, fig. 9). When they have passed through the wall they enlarge to the former size of the hypha, and grow on through the next cell, to be deflected as before upon reaching the opposite wall. On Pl. 5, fig. 9, a number of these attachment organs are represented, occurring in the wood of *Taxodium*, and also

---

* Miyoshi, Manabu. Die Durchbohrung von Membranen durch Pilzfäden. (Prings., Jahrb. f. wiss. Bot. 28 : 269. 1895).—Über Chemotropismus der Pilze. (Bot. Zg. 52 : 1. 1894.)

† Hartig, R. Zersetzungserscheinungen, etc. *pl. 8. fig. 11.*

‡ Brefeld, O. Untersuchungen über Schimmelpilze 4 : 112. *f. 11, 15.*

§ Marshall-Ward, H. On a lily disease. (Ann. Botany 2 : 319. 1889.)

in diseased wood of *Libocedrus decurrens*. At a one of the attachment organs has formed a branch. These organs adhere very firmly to the walls against which they are pressed; from the curved form of the hypha one is led to suppose that the pressure exerted by the hypha must be considerable. The reason for supposing these brown hyphae to be saprophytic is that they are usually found somewhere in the wood near a knot hole, where there is abundant opportunity for the entrance of saprophytes.

In many cases a form of mycelium with very thick walls occurs. This has few clamp connections and forms thick felts in the holes. This was found only in logs after they had been cut, so there is some reason for considering it as foreign to the disease. The great age of many of the cypress trees, and the consequent presence of numerous places where branches have been broken off, allows many fungi to get in which live on the dead and decaying wood, but which seem to have nothing to do with the peckiness. Their presence makes the study a difficult one at times, especially as they seem to grow rapidly and fructify readily. Thus a number of spore forms were met with, but in no case could these be brought into any connection with the colorless mycelium. One form was found very often (Pl. 5, fig. 5) also frequently present in diseased wood of *Libocedrus decurrens* and *Juniperus Virginiana*. The spores are almost round, about 1 μ in diameter, brown, with a distinct wall and a central shining body which is not affected by reagents. Many of the spores have short knobs. The spores occur in such numbers in the wood around the holes that it seems probable that they were formed in chains and may be considered chlamydospores. A number of times chains of two or three were found with fine remnants of hyphae attached. These spores were placed in cultures of dung, cypress agar, and gelatin, but have so far refused to germinate. It is possible that they represent some form of entophytic organism (*Chytridiaceae ? Phytomyxae?*) studied by

28

Fischer, Dangeard, De Wildeman and others. Brefeld *
records an instance of similar spores in *Peziza tube-
rosa*. These were constricted off in chains and refused
to germinate. Hartig † found spores in wood destroyed by
Polyporus sulphureus. These, he says, belong to some
saprophytic fungus, always found with *Polyporus sulphureus*. In wood of *Quercus alba* and *Q. nigra* destroyed by
Polyporus sulphureus, collected in New York and Arkansas,
similar spores were found, represented on Pl. 5, fig. 8, for
comparison. They seem to be constantly present wherever
Polyporus sulphureus has destroyed oak wood. The asso-
ciation of these two fungi is not understood as yet, and
awaits further investigation.

Besides the brown spores a number of others occur, a
few of which may be mentioned. One form, large, black
spores in chains, resembles Willkomm's ‡ *Xenodochus ligniperda* (Pl. 5, fig. 7). Another form, consisting of large
two-celled chlamydospores (Pl. 5, fig. 4), is not infre-
quent.

PROGRESS OF THE DISEASE.

In the early stages of the disease the wood turns yellow
in localized areas, about ¼ inch wide and extending longitudinally with the wood fibers for several inches (Pl. 1,
fig. 1). These areas are separated by intervening layers
of wood, unchanged in color. In the wood cells of the
yellow areas numerous hyphae of the colorless mycelium
are found. The larger hyphae extend longitudinally
through the cells and give off many branches which pass
and repass through the walls. The ultimate hairlike
branches reach every cell in the area. Numerous clamp
connections are to be seen. Between the yellow areas the
hyphae extend through the wood cells, passing through the

---

* Brefeld, O. Bot. Untersuchungen über Schimmelpilze 4 : 113.
† Hartig, R. Zersetzungserscheinungen, etc. 113. *pl. 14. f. 10-12.*
‡ Willkomm, M. Die mikroscopischen Feinde des Waldes. *pl. II.*
f. 8. 1866.

walls radially to another yellow area at that height or longitudinally to one above. Immediately around the yellow areas it looks as if the hyphae were passing through this wood as rapidly as possible. As the disease progresses the mycelium can be found only sparingly in the yellow areas and in the surrounding wood. Their former presence is indicated by the numerous holes in the walls.

From the facts presented, it seems that the growth of the fungus is about as follows: The mycelium starts at some point in the heart-wood where it flourishes in a limited area for some time. Some of the threads then grow out from this area (which is limited, for some reason or other), and grow both transversely and longitudinally from the original center. At points some distance from this center new centers are established, which in time are limited and form starting-points for further growth. One may cut through a young branch and find the cut surface perfectly sound. On splitting both pieces of the branch, one may find that at points several inches above the cut one or more distinct yellow areas are present, and the same may be true of the piece below the cut. In the wood between, numerous hyphae occur, which, however, do not spread in this wood. The areas where vigorous development has taken place ultimately become holes, and the tree then appears as already described, i. e., sound wood filled with lens-shaped cavities. The original hyphae are gradually absorbed, so that after a time the figure-8 holes in the walls are the only evidence of their former presence.

The path of the mycelium is always the shortest distance from hole to hole. This apparent avoidance of the wood between holes — an apparent preservation — is very striking. It is suggested that this is probably due to chemical influences which affect the hyphae in this manner. All attempts to grow the mycelium have so far failed. Media were prepared from decoctions of cypress wood and carefully titrated; they were then inoculated with fresh mycelium, which, however, did not grow. Fresh pecky wood has been kept in moist chambers now for almost three years without any sign of growth. Further experiments are in progress.

Propagation of Disease.

The constant presence of the colorless mycelium in diseased trees makes it seem probable that this is the vegetative part of a fungus which causes the decay. As has been said, no fruiting organ has yet been found, so the manner in which this disease is carried from tree to tree is still to be discovered. A large number of logs were split open, and in some of these, large places were occasionally met with where an old branch had been healed over, leaving a cavity. In this cavity dense white felts of the mycelium, in which numerous crystals of calcium oxalate were imbedded, were obtained. There was, however, no sign of a fruiting organ. In some boards beginnings of such felts were found but none of these have developed any further. Reasoning by analogy from the diseases of trees already known we ought to find at some time a pileus of some sort. That infection takes place through a broken branch or some part of the top of the tree is most probable. Many trees were cut down in which the "peck" could be traced directly to a broken branch, extending up and down from this point. This was especially marked where, as in a number of instances, the "peck" was confined to one side of a tree.

Localization of Disease.

The most characteristic feature in connection with this disease, distinguishing it from others so far described, is its peculiar localization, i. e., the destruction of the wood in distinctly localized areas. The formation of the holes has been described, and it has been noted that the contrast between diseased areas and sound wood is a marked one,

31

furthermore that fungus threads occur all through a given section of a tree.

The manner in which fungi influence their hosts varies considerably. One may consider the distribution of the mycelium within the host. There are but few references to this point in discussions on fungi. Tubeuf * says: "A large number of fungi have a mycelium which never extends beyond a very short distance round the point of first infection, and cause only local disease, frequently with no perceptible disturbing effect on the host. Such is the case with leafspot diseases." Thus Frank † describes the mycelium of *Gloeosporium Lindemuthianum* as causing a browning of the tissues as far as the mycelium extends. The same is true of *Cercospora*. The mycelium of *Aecidium Rhamni* on *Rhamnus frangula* has a local distribution,‡ so also that of many Erysipheae, for instance *Microsphaera densissima* also *Uncinula necator* of which an interesting case was recently described by Stevens.§ This localization of the mycelium may be due to mechanical obstructions, such as the veins of a leaf, as in *Puccinia Podophylli*, or to chemical reaction on the part of the host. The large majority of fungi have a mycelium which extends through large areas of their hosts. Wakker (l. c.) classifies parasitic fungi according to their effects on their hosts as producing either mechanical or chemical effects. By mechanical effects he understands such as are due to direct pressure. The vast majority affect their hosts chemically. Here again two classes may be distinguished, such as produce chemical effects "which will immediately, or otherwise exert a direct destructive influence

---

* Tubeuf, C. Freiherr von. Diseases of plants 16. (Eng. edit.)

† Frank, B. Über einige neue u. weniger bekannte Pflanzenkrankheiten. (Ber. d. deut. bot. Ges. 1 : 31. 1883.)

‡ Wakker, J. H. Untersuchungen über den Einfluss parasitischer Pilze auf ihre Nährpflanzen. (Prings., Jahrb. f. w. Bot. 24 : 505. 1892.)

§ Stevens, F. L. A peculiar case of spore distribution. (Bot. Gaz. 27 : 138. 1899.)

on their hosts and those which live for a longer or shorter period with their host without producing such effect." [*] To the first class belong all such plants as produce immediate death, like *Peronospora*, *Agaricus melleus* and many *Polyporei*, and those producing hypertrophies, such as *Gymnosporangium*, *Exoascus*, and others. To the second class belong many *Uredineae* and *Ustilagineae*, *Exobasidium*, etc. In the latter cases the mycelium may live for a long period in the cells without any perceptible effect on them. The reason for this " conservation " (Tubeuf, l. c.) is doubtless to be sought in complex chemical conditions which bring about one kind of effect with one, and another with a different fungus.

In all the cases just mentioned, one is dealing with living tissues capable of reaction of some sort. This reaction may take the form of starch accumulation, hypertrophied structures or the formation of products antagonistic to the growth of the invading fungus. The bacteria are a good example of organisms bringing about the last form of reaction, i. e., where the host produces substances which neutralize the poisonous products formed by the parasite. To what extent similar processes take place in plant cells is yet unknown, but there seems to be no reason why they should not.

In the heartwood of a tree one is dealing with a plant member to all intents and purposes dead, i. e., its power to react to any stimulus has been lost, so that such influences as would affect the distribution as well as chemical activities of a mycelium in a living member can have no bearing here. There is in the *Taxodium* a marked localization, and, as will be shown, this is also present in *Libocedrus decurrens*, *Juniperus Virginiana*, *J. Bermudiensis*, and to some extent in pines attacked by several of the *Polyporei*.

The localization of chemical action, for such the disinte-

---

[*] Tubeuf, C. Freiherr von.   Diseases of plants 21.

grating action on wood must be, cannot be due to mechanical causes, such as difference in the character of the wood cells, or the presence of obstructing layers. In the first place all wood cells are disintegrated, whether they be of the spring or summer wood, and the bounding lines of a cavity are not influenced by the harder summer wood as might be supposed. Then again no resistant layers, as such, excepting the harder summer wood, exist in the heartwood. The isolated resin cells seem to have no bearing in this connection. The only remaining explanation is to attribute the local decay to chemical influences, which prevent the decay from spreading beyond well-defined limits.

It is a well-known fact that certain kinds of wood are more durable, i. e., resist the destructive influences of fungi longer than others. Willows, for instance, are more easily destroyed than oaks or cedar. Frank * and Temme † have shown that in dicotyledonous trees a certain preservative gum is found. This is formed in all wounds open to the air, and occurs normally in all heartwood. This wound gum fills the vessels similarly to thylloses, and renders them impassable to air and water. The wound gum is insoluble in alcohol, ether, H_2SO_4, KOH, but soluble in hot HNO_3. These authors do not explain what causes the formation of this preservative material, beyond the fact that it forms when healthy wood is wounded and exposed to the air. In the Coniferae this substance is not present and Frank ‡ holds that the infiltration of resin takes the place of the gum. Hartig § finds a yellowish-brown mass in the cells adjacent to wounds in trees. This mass is usually much cracked. He calls it dried brown solution,

---

* Frank, B. Über die Gummibildung im Holze u. deren physiologische Bedeutung. (Ber d. deut. bot. Ges. 18 Juli, 1884).

† Temme, F. Über Schutz und Kernholz, seine Bildung u. seine physiologische Bedeutung. (Landw. Jahrb. 14 : 465. *Taf. 6, 7.* 1885.

‡ Frank, B. Die Krankheiten der Pflanzen 1 : 41. 1895.

§ Hartig, R. Zersetzungserscheinungen etc. 66. *pl. 11. fig. 7.*

and believes that it consists of decomposition products of wood exposed to the disintegrating influences of the outer air. These products are dissolved by water and penetrate far into the tree, bringing about the characteristic phenomena of wound rot. Frank * claims that Hartig has mistaken the nature of this substance, which he says is not a humus compound but wound gum, which acts as a preservative. A comparison of Hartig's figure and the one on Pl. 4, fig. 4, will show that in point of appearance the substance described by Hartig and the one in *Taxodium* cells are alike. I have also found such substances in wounds, and neither these nor the substance in *Taxodium* are the wood gum which Frank describes. I believe that Hartig is right when he calls them humus solution, but cannot agree that they are active in promoting decomposition. It might be added that Willkomm † ascribes the brown coloration of diseased pine wood to a humus compound which he says is formed from the cell-walls when they begin to decompose.

No substance corresponding to Frank's wound gum could be obtained from the *Taxodium*. An aqueous extract of the sound wood is yellowish in color, due to some coloring matter akin to curcumin. A number of analyses made of diseased wood failed to give any substances which might be regarded as preservative. The sole difference so far found between the normal wood and the diseased wood was the constant presence of the humus compounds described in the diseased wood.

There are numerous instances which illustrate the preservative and antiseptic properties of humus compounds. The preservative powers of peat deposits are well known. Peat is largely if not entirely composed of humus compounds of one kind or another. Its preservative and antiseptic prop-

---

* Frank, B. Krankheiten der Pflanzen 1 : 32. 1895.
† Die mikroscopischen Feinde des Waldes 68.

erties have been attributed to its humus acids. Thus Stutzer and Burri * killed cholera germs in a quarter of an hour with a decoction of peat. Lyell † speaks of the remains of animals and men, which had been perfectly preserved for many years in peat bogs. Kerner von Marilaun ‡ holds that the preservation of plant parts is brought about in moors by humus acids. "The dead plants are saturated with these acids and are not resolved into carbon dioxide, ammonia and water, but preserve their form and weight. The rapidity of decay varies inversely as the quantity of compounds of humus acids present." Also "the fact that fossil remains of Equisetums, Lycopodiums and Cycads . . . have reached us in such good condition, is explained by the presence of humus acids which are found so universally in peat.§ Ganong ‖ points out that the scarcity of nitrogen in peat bogs is due to the absence of bacteria "caused doubtless by an actively antiseptic quality of the bog water." Trees and stumps have often been found in bogs perfectly preserved. Lyell (l. c.) speaks of tree-trunks dug out of Irish bogs and used for masts, also of white cedar logs in New England bogs.¶ Other instances might be mentioned, but these will suffice to show that the humus compounds have antiseptic and preservative properties.

In the heartwood of the cypress one finds the wood substance being split up and destroyed. The decomposition stops after a time, and the fungus mycelium, which at

---

* Stutzer A., u R. Burri. Untersuchungen über die Einwirkung von Torfmull . . . auf die Abtötung der Cholerabakterien. (Zeits. f. Hyg. u. Inf. Krank. **14 : 453.** 1893.)

† Lyell, Sir Chas. Principles of geology 722.

‡ Kerner von Marilaun, A. The natural history of plants **1 : 262** (Eng. edit. by F. W. Oliver). 1894.

§ Kerner von Marilaun. l. c. **2 : 612.**

‖ Ganong, W. F. Upon raised peat bogs in the province of New Brunswick. (Trans. Roy. Soc. Canada ii. **3 : 131.** 1897.)

¶ Lyell, Sir Chas. A second visit to the United States 33. 1850.

first developed profusely, evidently stops growing. The threads become coated with a brown substance, which also fills many of the cells around the area where active decomposition has taken place, and saturates the cell walls. This humus substance is one of a class known to possess antiseptic properties. These facts suggest that the humus compound described above may in part be the agent which limits the disintegrating effects of the fungus.

ORIGIN OF THE HUMUS COMPOUND.

The origin of the humus compounds is still a matter of some uncertainty, owing to the intrinsic difficulties. Früh, who probably has paid more attention to this problem than any other investigator, says that we know as little about the successive stages which a plant member passes through, until peat is formed, as we do of the processes which bring about these changes.* The process is essentially a process of decay. It is at present recognized that decay may be due to chemical processes as such, distinguished from those brought about through the agency of living things. Where decay, or more properly a splitting up of highly complex organic compounds into simpler compounds such as carbon dioxide, ammonia and water, takes place without the aid of bacteria or fungi, it is largely a process of oxidation. If access of oxygen is prevented no decay takes place. Hartig,† speaks of the decomposition of plant members following death due to frost, as a process due to the action of oxygen on the dead organic substance; fungus mycelia get into the tissues after a time and hasten this decomposition. Früh‡ distinguishes two forms of decomposition not due to chemical changes per se (such as oxidation); these he calls " Gährung " and " Fermententwicklung."

---

* Früh, J. J. Über Torf u. Dopplerit 38.
† Hartig, R. Zersetzungserscheinungen, etc. 65.
‡ Früh, J. J. l. c. 39.

The first process is brought about " through the direct influence of the plasma of a living fungus, and is characterized by an evolution of heat and carbon dioxide. The other form is caused by a ferment excreted by living plants. This distinction can no longer be made to-day, as it seems probable that the first form of decomposition is also due to a ferment. Since humification takes place only under water, Früh holds that one might suppose a ferment the active agent in the formation of peat. But this cannot be true, for, if a ferment were the agent forming peat from vegetable substances, the process of humification would be a uniform one, that is, a given mass would be entirely transformed into peat. In a bog, however, this is not the case, for there are alternate layers, some of which are humified, others not. Früh, therefore, agrees with Einhof who says that " lack of free oxygen, a high degree of moisture and a low temperature, brought about by much moisture, bring about a decomposition of a peculiar kind, i. e., the formation of humus compounds or peat." He sums up as follows: * " The formation of peat is neither due to ' Gährung ' nor to a ferment but consists in the slow decomposition of plants, with the greatest possible exclusion of oxygen by water, and at low temperature. Bacteria have nothing to do with the formation of peat." This view of peat formation is the one generally accepted; thus, Shaler † explains it as due to the arrest of disintegration arising from the fact that the oxygen of the air does not have free contact with the carbon, and thus cannot convert it into CO_2.

This explanation practically states the fact that cellulose and lignin do change into a series of humus compounds, and that it is a process of chemical change. It does not explain what that change is and why it should take place.

---

* Früh. l. c. 49.

† Shaler, N. S. Peat deposits. (16th Rep. Director U. S. Geol. Survey. 4 : 308. 1895.)

A number of observers still maintain that fungi or bacteria are active in bringing about humification. Thus Höveler * finds that in the humus of a forest the mycelia of fungi initiate the process of humification. These mycelia are brown in color and are found in every humus soil. They belong to many different fungi and are characterized by possessing clamp connections. *Cladosporium humifaciens* Rostrop, he regards as the form most frequently present.

In decaying trees, and in such as are attacked by various fungi, such humus compounds are frequently present; they have been classed as decomposition products without further statement as to their origin. In decaying masses numerous fungi usually grow all through the mass, which makes it difficult to decide what the true humifying agent is. In the cypress a humus compound usually appears in the cells in which a definite fungus mycelium is growing. The same is true and probably more marked where the mycelium of *Trametes Pini* grows in pine wood. The latter turns red-brown very soon after the mycelium has entered the wood, and examination shows that this color is due to a humus compound. No humus compound is present in sound wood. This behavior of the compound makes it seem probable that the fungus in some way changes the cell-walls, and that the humus compound is one of the direct products of this change.

FERMENTS.

In the decomposition of wood it has been assumed that ferments take an active part. Enzymes which attack cellulose and lignified membranes are known. De Bary † and

---

* Höveler, W. Über die Verwerthung des Humus bei der Ernährung der chlorophyll-führenden Pflanzen. (Prings., Jahrb. für wiss. Bot. 24 : 290. 1892.)

† De Bary, A. Über einige Sclerotinien und Sclerotienkrankheiten. (Bot. Zg. 44 : 377. 1886.)

Marshall-Ward * isolated ferments which corroded cellulose membranes. Brown and Morris † discovered a ferment in germinating barley grains, and Vignal ‡ records a case of a ferment secreted by *Bacillus mesentericus*, dissociating vegetable cells by destroying the middle lamella. The action of the wood-destroying fungi is such, that Hartig and others have attributed the decay which they bring about to some enzyme excreted by the hyphae. That the same fungus produces several such enzymes must follow from the different effects which the same fungus has on the same wood. If then we assume such a cytohydrolytic enzyme to be formed by the *Taxodium* fungus, we find it destroying the wood about a certain center. As the mycelium grows along the vessels more readily than across them, a long hole is formed. As a result of the action of the fungus on the cell-walls, an acid humus compound is formed, which is deposited in the cells surrounding the center of fungus activity. It is not far to make the further assumption that after a time the amount of humus compound would be sufficiently great to stop the further development of the fungus in that area. The hyphae however pass through this area to a new center, where they begin over again. This would explain why the holes are approximately of the same size. The amount of antiseptic substance necessary to prevent further decomposition would be about the same in each area, and it would require the decomposition of a definite amount of wood to form this quantity. It may be objected that the holes are not always bounded by solid wood, but often run together. This would be explained by supposing the amount of humus formed at that point not sufficient to overcome the influence of the enzyme.

The conditions under which enzymes are active are

---

* Marshall-Ward, H. On a lily disease. (Ann. Bot. 2 : 346. 1888.)
† Brown and Morris. Jour. of Chem. Soc. 57 : 505. June, 1890.
‡ Vignal. Cont. à l'étude des bactériacées. (Thèse. Paris.)

variable. Marshall-Ward * finds that in a distinctly alkaline liquid the mycelium of *Botrytis* will no longer grow. The enzymes of most bacteria are effective only in alkaline or neutral media † while those of many fungi are active in distinctly acid media although growth is more vigorous in neutral solutions.‡ Green § notes " the possible significance of the inhibitory effects of traces of acid or alkali in the solution in which the enzyme is working," and Smith ‖ has made similar observations. It is suggested that the humus compound may, because of its acidity, bring about conditions unfavorable to the activity of the enzyme formed in the cypress wood. As the humus compound is at first in liquid form, it saturates the wood cells for some distance around the hole, and thus completely fills the space where the mycelium is growing and many of the cells outside of this space. This explains why the hyphae as a rule grow out from the holes in straight lines without branching much in the wood surrounding the holes.

The enzymes are usually soluble in cold water, and can be precipitated from a solution by an excess of alcohol. Masses of diseased wood finely divided, as well as masses of mycelium, were digested with cold water for 27 hours; then to four parts of alcohol one part of the extract was added. A gray flocculent precipitate was obtained which on drying in a vacuum turned slightly darker. It was slightly soluble in water. Sections of *Taxodium* wood, young bean stems, etc., when placed in such a solution showed no perceptible change. As Hansen (l. c.) pointed

* Marshall-Ward, H. On a lily disease. (Ann. Bot. 2 : 319. 1889.)

† Fermi, Claudio. Weitere Untersuchungen über die typischen Enzyme der Micro-organismen. (Cent. f. Bact. u. Parasitenkunde 10 : 404. 1891.)

‡ Hansen, A. Die Verflüssigung der Gelatine durch Schimmelpilze. (Flora n. s. 47 : 88. 1889.)

§ Green, J. R. On Vegetable ferments. (Ann. Bot. 7 : 83. 1893.)

‖ Smith, E. F. Sensitiveness of certain parasites to the acid juices of host plants. (Abstract in Bot. Gaz. 27 : 124. 1899.)

out, this method of separating enzymes is very unsatisfactory, as it weakens the enzyme and may even destroy it.   In this case it is probable that much of the precipitate consisted of soluble humus compounds, and as these are likewise precipitated by alcohol a separation becomes difficult.

As the humus compound is insoluble in water (except a minute trace) it is difficult to add it to any culture media. It was dissolved in very weak KOH and added to agar and bouillon tubes which were inoculated with various bacteria and fungi.   To a similar series of agar and bouillon tubes the KOH solution was added and likewise inoculated.   In this double series no additional inhibitory effects due to the humus compound were evident.

The conclusions arrived at in this chapter indicate that the humus compound found in the wood surrounding the holes is formed because of the action of a fungus on the cell-walls of the wood, and that it is probably one of the products effective in preventing the unlimited spread and destructive action of the disintegrating powers of that fungus.

AGE OF THE FUNGUS.

Taxodium distichum is an interesting tree in that it is one of the surviving members of a race of trees which were prominent in geologic times.   Any disease which it is affected with may possibly have come down to the present day with its host.   But few fungi are known in fossil condition.   Unger * describes mycelia from the wood of a Tertiary tree; Williamson † figures a fungus, *Peronosporites antiquarius* from a stem of *Lepidodendron* (the same is also found in coal beds).   Conwentz ‡ found a mycelium in fossil wood of *Rhizocupressinoxylon unira-*

---

* Unger, F. Chloris protogaea. 1847.

† Williamson, W. C. On the organization of fossil plants of the coal measures. — Calamites. (Phil. Trans. R. S. L. **161**:477. 1871.)

‡ Conwentz, H. Fossile Hölzer von Karlsdorf am Zobten 27. Danzig. 1880.

diatum. The mycelium had clamp connections and swellings like those of *Agaricus melleus.*

In accounts given of *Taxodium* logs found buried at various points * no mention is made of any defect. While in southern Louisiana last winter a number of sections of buried *Taxodium* logs were obtained.† These were found several miles back from the Mississippi river at an average depth of 10 ft. below Gulf level. Compared with other cypress logs found, these are not very old, but they are sufficiently far removed from the present time to deserve notice. In two of these logs unmistakable signs of peckiness were found. There was very little mycelium, but a sufficient number of hyphae with clamp connections were seen to justify the conclusion that they were the same as those growing to-day. The holes were few in number, but were not to be mistaken. It is probable that if other and older logs were examined more instances would be found. It would seem therefore that the disease is one which has extended back for some thousand years at least, and probably further.

---

* Lyell, Sir Chas. Travels in N. A. in the years 1841-2. **1 : 114.** — He says of cypress logs buried in the Dismal Swamp: "When thrown down they are covered by water, and never decompose except the sap."

Lyell, Sir Chas. A second visit to the U. S. 249. (1850). — (Cypress buried at the mouth of the Altamaha river.)

Carpenter, Wm. Account of the bituminization of wood in the human era, etc. (Am. Journ. of Sc. & Arts **36 : 118.** 1839). — (Buried cypress forest at Port Hudson, La.)

Bartram, W. Travels through North & South Carolina, Georgia, E. & W. Florida 66. London. 1792.

† The logs were found at the following points : —

No. 1. Standing stump 9 ft. below surface, 7850 ft. back from river on Jourdan ave. *No. 2.* Horizontal log, butt 30 inches; center 10 ft. below surface, 8325 ft. from river on Jourdan ave. *Nos. 3 & 4.* Standing stumps same locality as No. 1. No. 3 had 260 rings in the heart wood. The surface where the samples were taken reads about 21 Cairo datum, mean Gulf level, 21.26 C. D., i. e. they were therefore about 10 ft. below Gulf level. See also Chart No. 76 of the Mississippi River Commission, for location of Jourdan ave.

43

If one considers the manner in which a fungus disease attacks plants at the present day, one will find that closely related plants are apt to be afflicted by the same disease. Thus *Plasmopara Cubensis* grows on a large number of genera of the *Cucurbitaceae; Gymnosporangium Nidus Avis* (*Aecidium*) on several genera of the *Pomeae* (*Rosaceae*); *Trametes Pini* on several genera of the *Coniferae*, and so on. Judging by analogy, one might expect genera nearly related to *Taxodium* to be diseased similarly to *Taxodium.* There are but a few genera, closely related to *Taxodium*, which grow at the present time. In North America: *Taxodium mucronatum* is found in Mexico; *Sequoia gigantea* and *S. sempervirens*, in California; *Libocedrus decurrens*, in California and Oregon; and the less closely related species of *Juniperus.* In addition to these there are a number of other species scattered over the globe, thus *Libocedrus Douiana* and *L. Bidwillii* in New Zealand, *Libocedrus cupressoides* and *L. Chilensis* in Chili, also a dwarf species in Iceland. A closely related tree, *Glyptostrobus Europaeus*, is found in some of the southeastern provinces of China. All of these trees were common over the whole earth in Tertiary times, and if a disease was common to them then, one might expect to find that to-day. Of the species enumerated, the *Sequoias* are apparently free from diseases of the wood * while *Libocedrus decurrens* and the species of *Juniperus* so far seen, are affected by diseases which cause local rotting of the wood much like that of the cypress. Of the other trees nothing is known so far. The fungus which causes the decay in *Libocedrus* is described in the following, while that found in the trunks of *Juniperus* species is to be described in a separate paper soon to appear.

---

* Sargent, C. S. Silva of North America **10** : 140. 1896.

THE " PIN " DISEASE OF LIBOCEDRUS DECURRENS.

HISTORICAL.

In 1879 Harkness * published a note in which he called attention to a peculiar rot which occurs in the heart wood of the incense cedar, *Libocedrus decurrens.* Mayr †️ mentions what is evidently the same disease and describes it as follows : " A certain fungus, *Daedalea vorax,* appears to be very destructive, destroying the heart wood of standing trees ; the fungus colors the wood red-brown and forms large lens-shaped cavities, at the same time the wood becomes very " brüchig." In 1896 the following appeared in Sargent's Silva : ‡️ " The trunks of *Libocedrus decurrens* are frequently honey-combed and its value destroyed as a timber tree by *Daedalea vorax,* which destroys rounded masses of wood, disposed in long rows, sometimes extending through the length of the trunk, reducing them to a cinder-like powder." In the note published by Harkness one is led to believe that *Daedalea vorax* attacks *Abies Douglasii,* — not *Libocedrus. Daedalea vorax* is reported as growing on *Libocedrus decurrens* by Harkness.§ In a letter received from Dr. Harkness last year he says : " *Daedalea vorax* is a fungus which causes the rot in *Abies Douglasii,*" etc. As to the *Libocedrus* disease he says : " Nothing could be found except mycelium which permeates through the diseased portion. No visible signs of any spores were seen. A careful search fails to reveal any of the fungus either among the roots or the surface of the tree, nothing indeed to indicate its presence until the tree has been felled." He says furthermore that the note

---

* Harkness, H. W. A foe to the lumberman. (Pacific Rural Press. Jan. 25, 1879.)

†️ Mayr, Heinrich. Die Waldungen von Nordamerika 324. München. 1890.

‡️ Sargent, C. S. Silva of North America 10 : 134. 1896.

§ Harkness & Moore. Cat. of Pacific Coast fungi 12. (Read before the Cal. Acad. of Sciences, Feb. 2, 1880).

in the catalogue of Pacific Coast fungi which records *Daedalea vorax* on *Libocedrus* (l. c.) is an error, and that instead of *Libocedrus* it ought to read *Abies Douglasii*. Mayr's statement is therefore the only one ascribing this disease of *Libocedrus* to *Daedalea vorax*, for the note in the Silva was based on the statements of Harkness and Mayr. In view of the fact that Mayr's report has never been confirmed I am inclined to the belief that *Daedalea vorax* has nothing to do with the decay of *Libocedrus*. This would leave the identity of the fungus which is responsible for this trouble as obscure as in the case of *Taxodium distichum*.

CHARACTER OF THE DISEASE.

Specimens of diseased wood received from various parts of California and Oregon have the appearance shown in Plate 2. The heartwood is full of lens-shaped cavities filled with a very brittle brown material. The latter is evidently the wood which formerly filled the cavity, but has been changed and has shrunken considerably. The cavities are placed irregularly in the wood with their longest diameter parallel to the wood cells. They vary considerably in size, from 1 inch long and $\frac{1}{4}$ inch wide to 10 inches long and $1\frac{1}{2}$ inches wide. In the majority of cases the separate cavities do not communicate with one another, but occasionally they do, as is evident from the cavities at the right side of the figure. The line of demarkation between sound wood and the brown decayed wood is a very sharp one. When the decayed wood is removed, the cavities have a sharply-defined, smooth bounding surface. In most respects the appearance of the wood is like that of diseased *Taxodium* wood.

OCCURRENCE.

Concerning the prevalence and mode of occurrence of this disease, only such facts can be given as were learned from correspondents — particularly from Dr. Harkness,

46

Mr. A. J. Johnson, and a number of lumber companies in California, Oregon, and Washington. The disease is one which resembles the one in the cypress in its method of growth. The decay begins somewhere in the upper part of a tree and proceeds both up and down, the lens-shaped cavities appearing at first as darker areas in the wood. Older trees are very liable to be diseased. One correspondent, from southern Washington, says: " The proportion of trees affected is very large. We might almost say that the trees are generally so affected in this country." From Placer Co., Cal., another correspondent writes: " Probably more than one-half of the trees are affected in a greater or less degree." From intermediate points similar reports have been received. Young trees, i. e., such as are under 12 inches in diameter, are not apt to be seriously diseased. Climatic and soil conditions seem to have as little influence on the prevalence of the disease, as they do in the case of pecky cypress. Wherever *Libocedrus decurrens* grows, the defect is also to be found, i. e., from central California northward, as far as it has been possible to learn. The diseased wood is quite durable and can be used for fence posts, scantling, or for wood sills in buildings. The diseased wood is sold for $1–$3 less than sound cedar, per thousand ft. B. M., depending upon the degree of decay. This is an indication that it is at least capable of being used for some purposes. It might be mentioned here that boards cut from trees of *Juniperus Virginiana* affected with a similar disease were recently pulled off a barn where they had been 52 years. The Stimson Mill Co., of Ballard, Wash., writes: " We do not make any difference between sound and rotten cedar; $8 is the price for cedar delivered."

NAME.

The only name which has been learned which is applied to this disease is " pin rot." The term " pecky " has been applied to a form of decay in the cypress in which the wood is destroyed in local pockets. As this is a distinct form of wood destruction I would apply the term " pecky " to all forms of destruction where pockets or holes are formed as in the cypress. One would therefore call the affected *Libocedrus* wood " pecky cedar."

STRUCTURE OF DISEASED WOOD.

The normal wood of *Libocedrus* differs but little from that of *Taxodium distichum*. Penhallow * places the two genera side by side. The diseased wood is decidedly different from the healthy wood. It has the appearance of a brown charcoal, breaks with a dull fracture and when pressed crumbles into a fine powder. In the mortar an impalpable dust is formed. In this respect it is very different from much-rotted cypress wood. In the latter the chemical transformation is far from uniform. Diseased *Libocedrus* wood is changed throughout; both the spring and the summer wood are changed, and very rapidly at that, i. e., there are no intervening steps as in *Taxodium*. A section made through the edge of a diseased pocket shows that at a certain point the cells are brown (Pl. 4, fig. 2). It will be noted that the color of fig. 2 is the normal color of the wood. Hand in hand with this coloration goes a shrinkage of the middle lamella, so that the walls of the tracheids become much thinner. They have lost all tenacity. If a piece of charred wood is boiled in water for a few moments it can be pressed into any shape like a piece of dough. Sections on a slide can be pushed about so that the cells assume a rhomboidal shape, i. e., the whole acts like a network of fine flexible wire. This is to some extent visible in Pl. 4, fig. 2, where a number of the walls are much bent, and do not have the rigid appearance of the healthy wood

---

* Penhallow, D. P. Generic characters of N. A. Taxaceae & Coniferae. (Trans. Roy. Soc. Canada II. 2: 51. 1896.)

(fig. 1). The shrinkage of the wall causes breaks to appear in the pits (Pl. 5, fig. 11) and after a time in the walls (e). The shrinkage in a large mass of wood after a time becomes so great that the wood breaks at some point and gives rise to the appearance to be noted in the long hole at the left of the block in Plate 2. The three lamellae of the wood-cell are distinct even in greatly "charred" wood (Pl. 4, fig. 2).

The chemical nature of the wood cells has been entirely changed, and, as has been said, the change from sound wood to completely charred wood is immediate so far as microchemical tests can show. With dilute KOH the diseased wood swells to two or three times its size and the breaks in the walls close. A large per cent. is soluble in KOH and from such a solution humus compounds similar to those found in the cypress are obtained. Chlor-iodide of zinc turns the walls brown. When treated with dilute nitric acid the secondary lamellae gradually dissolve and there is left a skeleton framework composed of the primary lamella, the intercellular substance at the angle, and the fine membrane of the pits with the thickened torus. The solution takes place very gradually and can be followed very readily in a thin section. The nitric acid evidently dissolves out the substances into which the secondary lamella has been changed, leaving the more resistant primary lamella intact. From the HNO_3 solution a heavy flocculent orange mass is precipitated when excess of water is added. This precipitate is very soluble in alcohol and acetic acid, slightly so in ammonia, insoluble in ether, chloroform, benzine or acids. When dissolved in absolute alcohol, and cooled, no crystals form, but an oily substance settles on the walls of the dish as the alcohol evaporates. No further attempt was made to determine what this is. Nitric acid and potassium chlorate dissolve the entire wood substance. With H_2SO_4 the walls turn black and swell considerably. Phloroglucin and hydrochloric acid stain the rotted wood carmine red verging toward orange, indicating the presence of coniferin. When treated for twelve hours with Javelle water and then stained with chlor-iodide of zinc the primary lamella turns light brown; with methylen blue it stains deep blue, indicating the presence of pectic substances.* The skeleton framework obtained after treatment with nitric acid stains blue with cellulose stains. This behavior towards various reagents shows that most of the cellulose has been removed and that the lignin substances have been transformed into substances readily soluble in nitric acid.

A number of chemical analyses were made of charred wood, following the method given by Allen † for determining the compounds found in wood. The wood was finely rasped and pulverized and dried at 100° C. After an aqueous extraction, the wood was extracted with alcohol and then with ether. 8.33% was found soluble in alcohol. The dried residue was hard, and broke with a bright fracture. It had all the attributes of a resin. Small quantities of pectic substances were found present, and a number of other products which were not determined. The rotted wood does not restore polarized light.

Wood Between the Holes.

The wood between the rotted areas is, as in *Taxodium*, perfectly sound as far as its structure is concerned. It reacts with reagents similarly to healthy wood. In the cells immediately surrounding the diseased spots, especially in the wood parenchyma and medullary rays, a red-brown substance is always present, which fills the cells as with plugs. It is very resistant toward acids and while boiling nitric acid dissolves the wood it does not affect this substance. Oxalic acid turns it black very quickly, also potassium bichromate and iron salts. These reactions

---

* Mangin, A. Sur la présence des composés pectiques dans les végétaux. (Comptes rend. etc. 109 : 577. 1889.)

† Allen, A. A. Commercial organic analysis 1 : 323. 1898.

would place it with the tannins. Hartig found tannin in decayed wood, whereas it was not present in sound wood, and in the present case there seems to be a similar instance. What the origin of the tannin may be I do not venture to say.

Aside from the tannin a brown humus compound, similar to that found in *Taxodium*, occurs. It is found in the form of irregular granular masses which readily dissolve in dilute KOH. The medullary rays in particular are filled with this substance (Pl. 4, fig. 2); it seems to permeate the cell-walls, for these turn the characteristic yellowish-brown color on addition of KOH, and the tracheids become filled with the brown liquid. Extractions of the surrounding wood with KOH yield considerable quantities of the compound. Nowhere were any dried plates found, such as were described for the cypress.

MYCELIUM AND SPORES.

The mycelium found in the diseased *Libocedrus* wood agrees so closely in appearance with that found in the *Taxodium* that the drawing on Pl. 5, fig. 1 may represent it as well. Few hyphae are to be found in the charred wood or the wood about the holes. Abundant evidence of their having been present is seen in the numerous holes which puncture the walls of the charred wood in all directions (Pl. 4, fig. 2). No preference is shown for the pits. The hyphae are most abundant in wood away from the rotted holes. They are colorless, branch frequently and are provided with a large number of clamp connections. The finest threads pass through the walls in all directions. Between the rotted areas the hyphae usually extend directly from hole to hole, just as in the *Taxodium*. In places the mycelium collects in large masses or felts; in these felts the hyphae are matted. Many crystals of calcium oxalate give the whole a white appearance.

A brown mycelium like that found in the cypress was

51

found in a number of cases. How common this is cannot be said, as the number of specimens examined was from but a small number of trees. The threads have marked attachment organs (Pl. 5, fig. 9 " d–f ") which have been described under the cypress disease.

In the rotted wood, and particularly around the same, the cells are often filled with great masses of spores like those seen in isolated cases in the cypress (Pl. 5, fig. 2). These spores are present in such numbers, that they often completely fill the tracheids. Several spores were found with very fine hyphae attached (fig. 5) and many showed small knobs at one end. It will be necessary to see a large number of trees to determine where these spores came from.

LOCALIZATION.

The localization of the diseased areas is quite as marked in *Libocedrus* as it is in *Taxodium*. One may have a block of wood 3×3×1 in. which looks perfectly sound, but when split longitudinally it may contain a sharply defined lenticular hole. It is suggested that probably similar reasons to those given for the cypress hold here. The investigation with respect to this point is to be regarded as but begun. When it becomes possible to grow the fungus found in the holes one may expect to reach more decisive conclusions.

SUMMARY.

In the foregoing, two forms of decay have been described, one destroying wood of *Taxodium distichum*, the other of *Libocedrus decurrens*. In both cases the wood is destroyed in localized areas, which are surrounded by apparently sound wood. The cell-walls are changed into compounds which diffuse through the walls and fill the cells surrounding the decayed center; and these have been called humus compounds. In both, a fungus mycelium occurs with strongly marked characteristics, which flourishes

52

within the diseased centers and grows between these centers without affecting the intervening wood. This wood can be utilized for many purposes even when much rotted, and in neither case does the mycelium grow after the tree has once been cut down. The two trees thus diseased, both representatives of a race of trees the majority of which are extinct, are closely related genetically, although growing in different parts of the country. The two forms of decay differ but slightly, and not more than might be expected in two woods of different character. Taking those facts into consideration, it appears probable that the two diseases are caused by one and the same fungus, the fruiting form of which has not yet been found.

EXPLANATION OF PLATES ILLUSTRATING DISEASES OF TAXODIUM AND LIBOCEDRUS.

Plates 1, 2, plate 4, fig. 3, and the coloring of plate 3, fig. 2, and plate 4, fig. 2, were prepared under my direction by Miss Harriet P. Learned.

Plate 1. — 1, Branch of *Taxodium distichum*, showing early stage of the pecky disease. The wood turns yellow in longitudinal lines ($\times \frac{1}{3}$). 2, A block of *Taxodium distichum* cut from the heart of a tree several hundred years old, showing advanced stage of peckiness. In a large number of trees the rotted portion is more yellow than that shown in the figure ($\times 1$).

Plate 2. — A block of *Libocedrus decurrens* showing advanced stage of the pecky disease. The rotted wood has fallen out from the holes at the right of the figure leaving a smooth surface ($\times \frac{1}{3}$).

Plate 3. — 1, Transection of pecky cypress wood. The section was made so as to include some of the much rotted wood, seen at the bottom of the figure, also some of the sound wood. It was stained with phloroglucin and HCl. The violet of the original section was somewhat more marked than is the color in the figure. The portions staining violet indicate wood which has not been affected by the fungus, those staining yellow show where the coniferin elements have been extracted: 'm' medullary rays; 'k' cell-walls from which the coniferin has been extracted; 'p' normal cell-wall; 'd' primary lamella resisting the disintegrating factor longer than the secondary lamellae; 'h' perforation of cell-wall made by fungus hypha (magnification same as fig. 2). 2, Transection of pecky cypress wood, showing the transition from sound wood to humus compound, after staining with phloroglucin and HCl.: 'g' primary lamella, unaffected; 'p' small masses of humus compound resulting apparently from the transformation of the tertiary lamella; 'i' wood staining yellow, an intermediate stage between the sound wood and the humus compound; 'h' a thicker layer of humus compound than the one indicated at 'p'; 'c' a still more advanced stage in the humus formation; 'e' the entire cell-wall has been transformed into the humus compound; 'u' piece of cell-wall not yet changed to humus compound.

Plate 4. — 1, Transection of sound wood of *Libocedrus decurrens*, showing spring and summer wood: 'h' brown hypha with attachment organs; 's' spores often found in the wood cells. 2, Transection of diseased wood of *Libocedrus decurrens*, i. e. wood from one of the pockets. The color is the natural color of the wood. The medullary ray is filled with brown humus solution. 3, Block of *Taxodium distichum* showing pecky hole lined with white fibers, consisting of pure cellulose ($\times \frac{1}{2}$). 4, Two tracheids from wood surrounding a diseased spot in *Taxodium distichum*. The tracheids are filled with brown humus compound which has cracked in drying.

Plate 5. — 1, Mycelium from decayed wood of *Taxodium distichum*, showing the numerous clamp connections. 2, Spores from pecky wood of *Libocedrus decurrens*. (The line at the top is 10μ.) 3, Portion of a tracheid near diseased area of *Taxodium distichum*. The pits appear corroded because of a peculiar arrangement of resin globules. (Magnification same as fig. 2.) 4, Brown chlamydospores from rotted wood of *Taxodium distichum*. 5, Brown spores from wood of *Taxodium distichum*. These are like the ones found in the red cedar. 6, Spores from wood of *Quercus alba* destroyed by *Polyporus sulphureus* (from Williamsville, Mo.; magnification same as fig 5). 7, Spores from wood of *Taxodium distichum*, resembling Willkomm's *Xenodochus ligniperda*. 8, Minute bodies, which appear when the humus compound is slowly dissolved away. Two views are represented (magnification same as fig. 2). 9, Mycelium showing attachment organs: 'a–c' from wood of *Taxodium distichum*; 'd–f' from wood of *Libocedrus decurrens*. 10, Longisection of pecky cypress wood, showing gradual disintegration of the tracheids: 'a' normal tracheid filled with humus compound; 'b' similar tracheid with colorless mycelium; 'c' tracheid with pits looking as if corroded; 'd' tracheids with walls which are beginning to contract; 'e' tracheid in which the walls show spiral cracks; 'f' and 'g' tracheids showing final stages in the process of solution. (Magnification same as fig. 1.) 11, Longitudinal section through pecky wood of *Libocedrus decurrens*: 'a' normal tracheid; 'b' tracheid showing beginning of disintegration, the pits show cracks, some spores are collected near a wall; 'c' and 'd' tracheids which have contracted considerably, showing cracks in the pits and the wall. (Magnification the same as the preceding figure.)

54

DISEASES OF TAXODIUM AND LIBOCEDRUS.

Plate 6. — Upper figure a pile of pecky cypress boards at Lutcher, La. The boards have been exposed some time, so that the rotted wood has been washed from the holes. The lower figure is a photograph of the vertical banks of a ditch on Dauphin St., Mobile, Ala. (in front of the house of Dr. Chas. Mohr). The bank is lined with pecky cypress boards, which are held in place by horizontal braces.

55

1

2

PECKY CYPRESS.

PLATE 2.

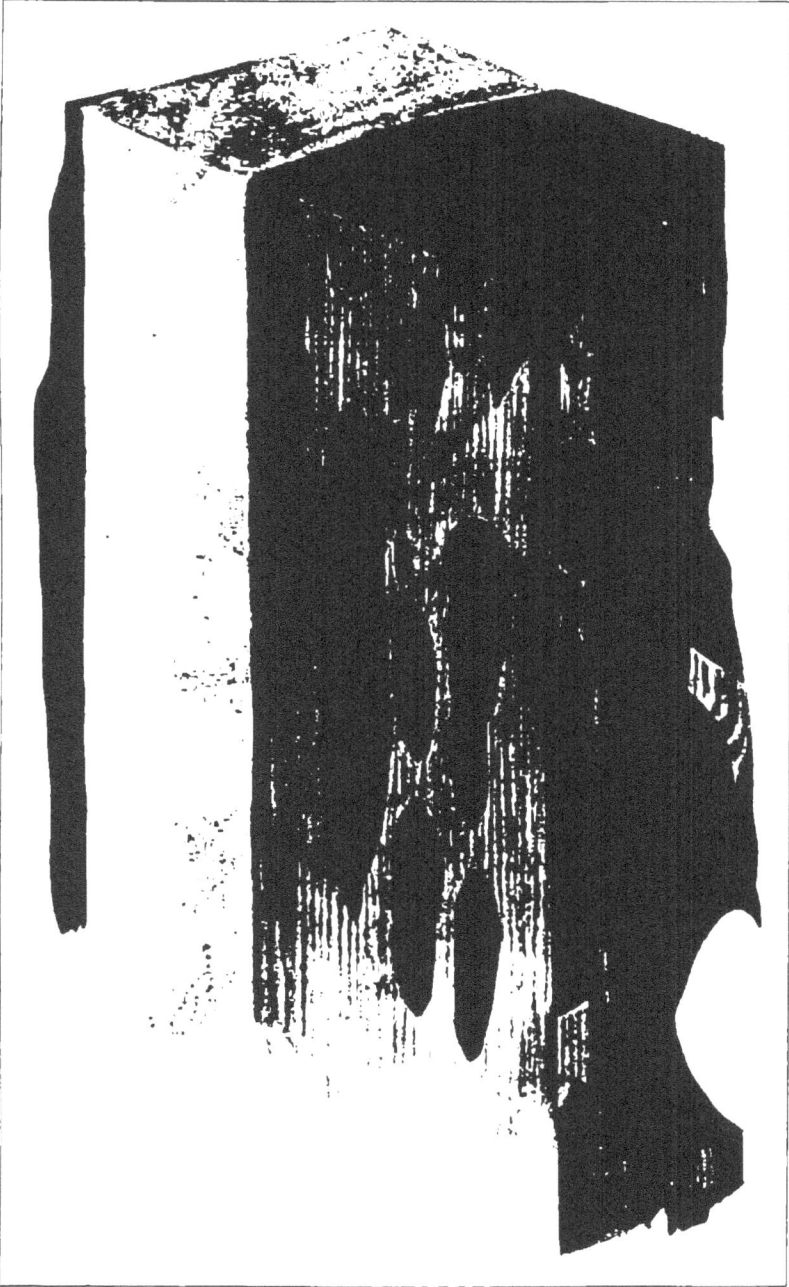

PECKY INCENSE CEDAR.

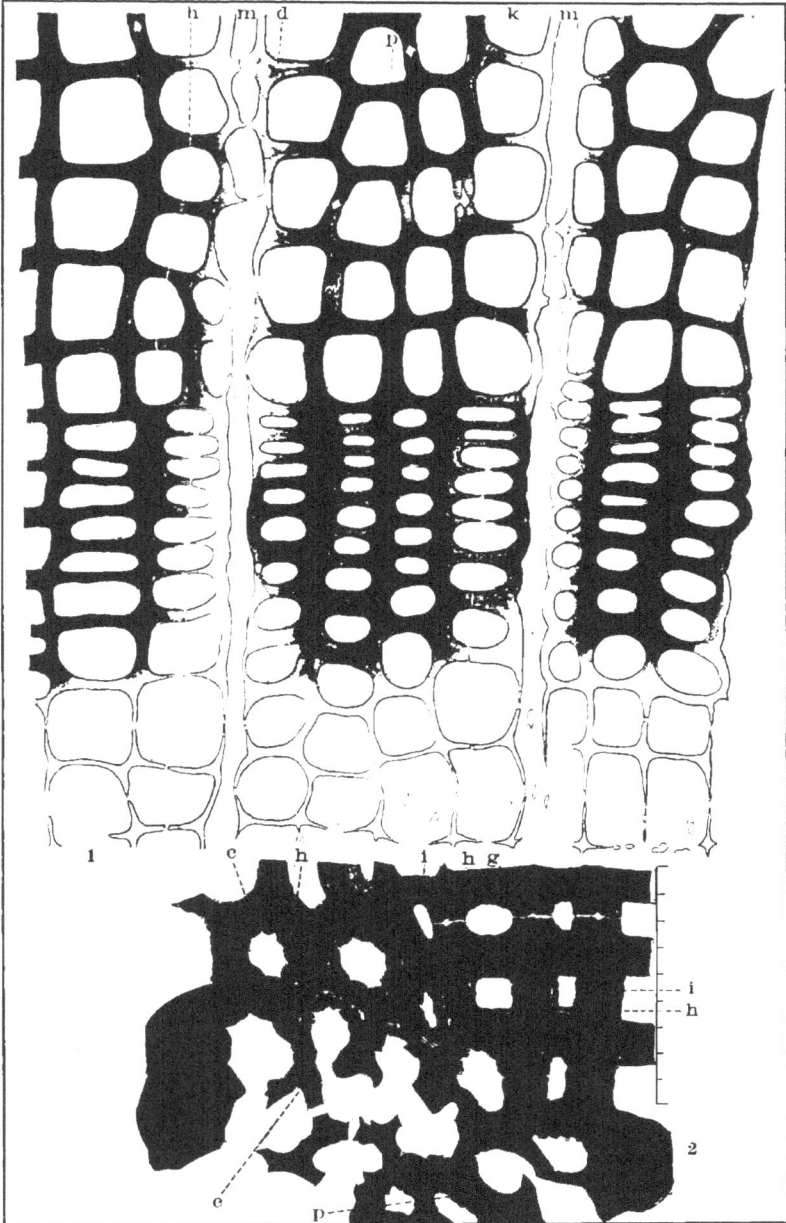

PECKY CYPRESS.

PLATE 4.

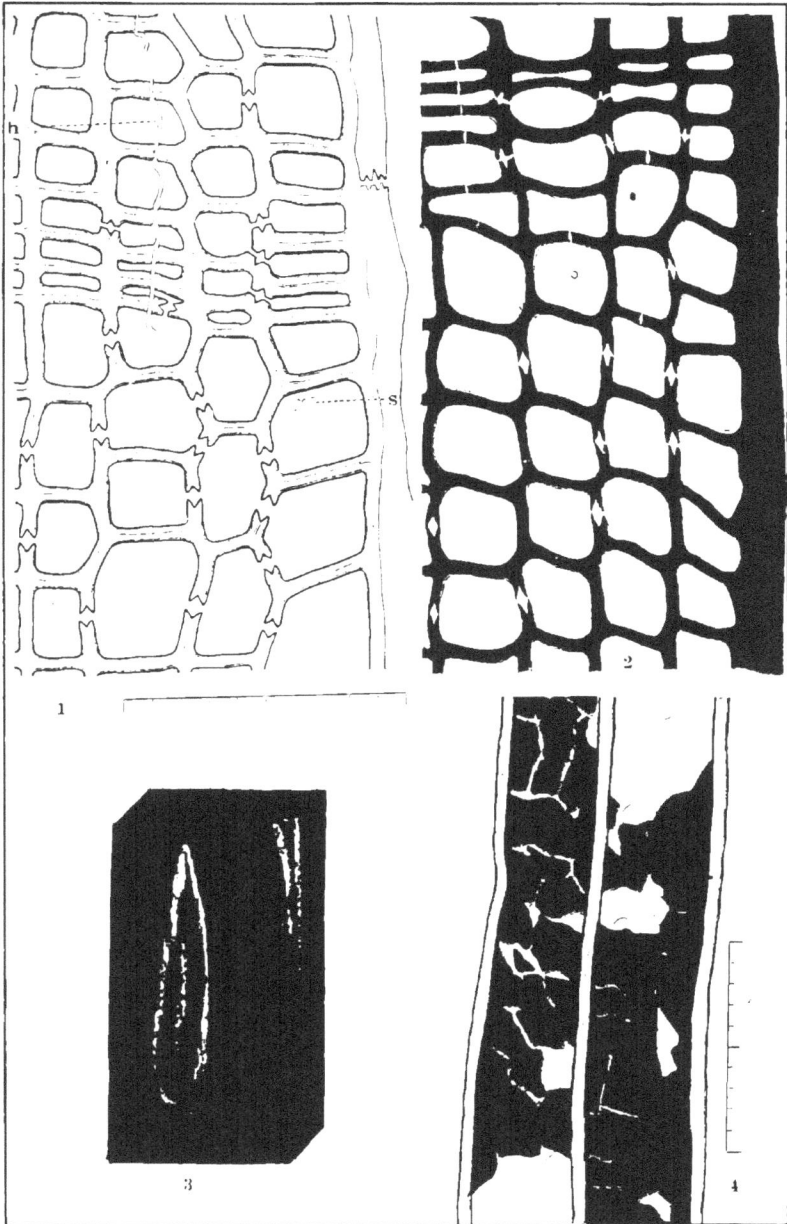

PECKY CYPRESS AND CEDAR.

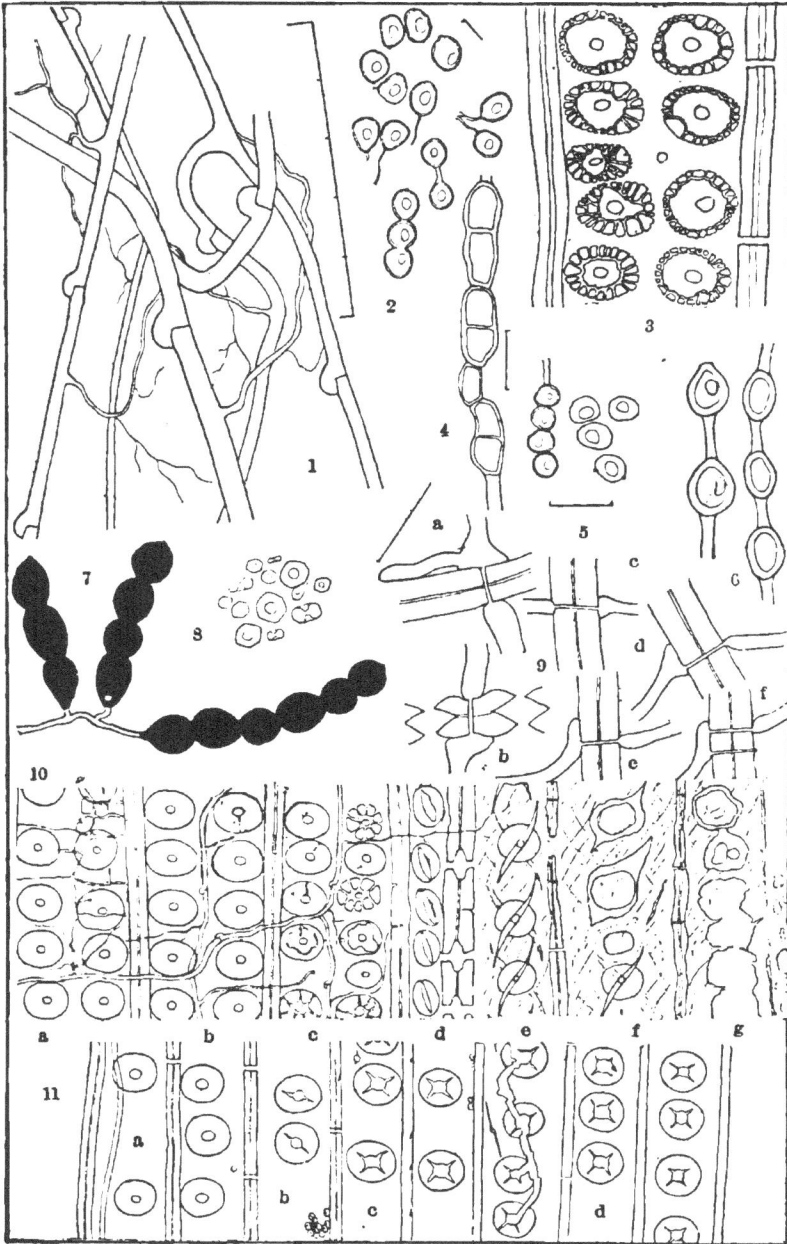

PECKY CYPRESS AND CEDAR.

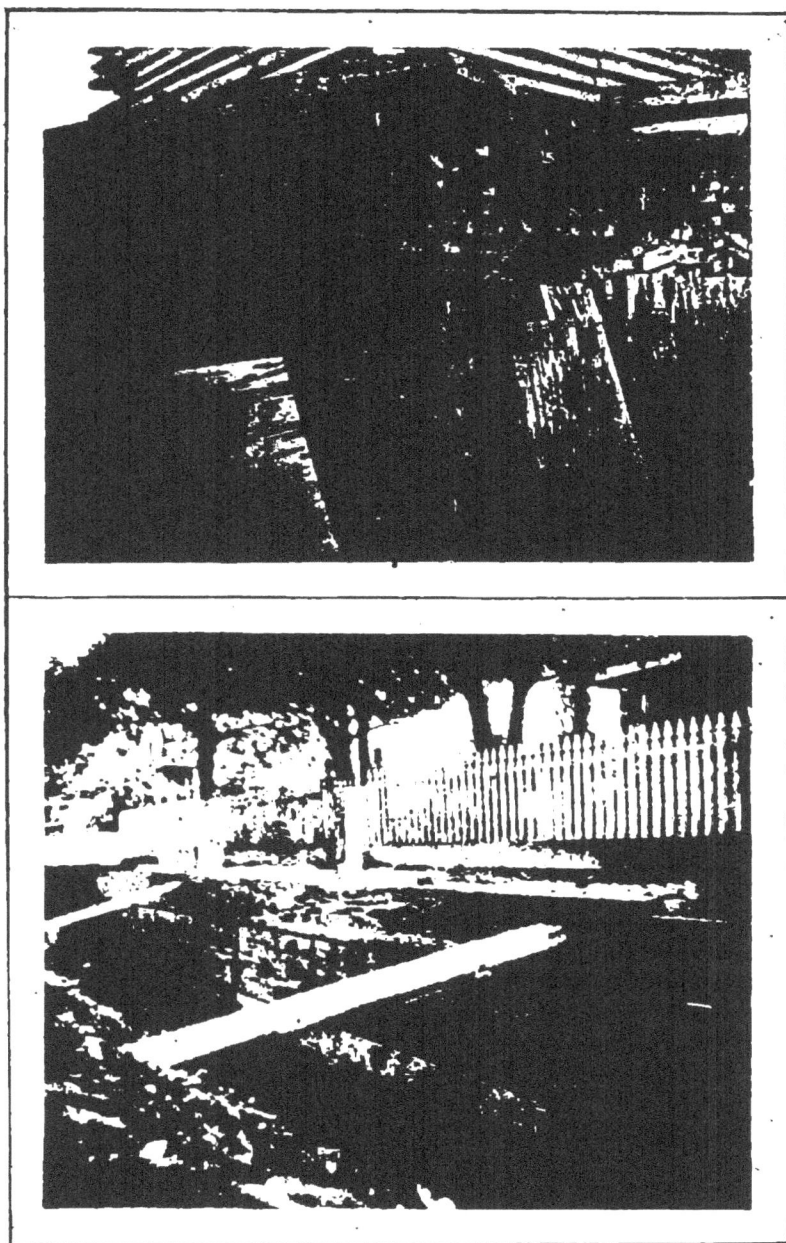

PECKY CYPRESS.

9783744716536